Somatomotor Increscence on Prodigiosity of the Athlore Striketics Physiants

G Baruch

Somatomotor Increscence on Prodigiosity of the Athlore Striketics Physiants

First Edition December 2025

Wriiten by G Baruch

CONTENTS

LIST OF TABLES

LIST OF CHART

CHAPTER-I
INTRODUCTION

1.1 Introduction -

Games and sports have been part of human life since ancient times. In the early days, they were essential for survival—such as hunting for food, seeking shelter, and protecting oneself from wild animals or enemies. Over time, these activities evolved into forms of recreation and gradually became an integral part of human culture. Today, games and sports serve as a powerful unifying force and contribute greatly to national and international integration. Sport is a way through which we use our physical abilities to play and stay active. It is crucial in many ways: when the body functions better, the mind also becomes sharper, as the two are closely connected.

Sports help us relieve stress, momentarily forget our problems, and enjoy ourselves despite the pressures of daily life. Sport is both recreation and competition, offering individuals opportunities to use their leisure time in meaningful and refreshing ways. Sport also carries strong cultural significance. It plays an important role in people's daily lives and brings communities together. For many, sport is central to their sense of identity, giving them a focus, a circle of friends with shared interests, and a structure around which they can organize their time and lifestyle.

Globally, sport is valued for promoting physical health and overall well-being and is supported by educational, social, and health systems. At any age, engaging in physical and sporting activities improves fitness, sociability, and provides an outlet for natural competitiveness. In this way, sport serves a wide range of important functions in human life.

From the school child throwing their first ball as a part of an learning progression, thru to the multi-million dollar pitcher on a major league baseball team, sport is an important awareness for, and a significant a part of, the society wherein we stay (Chandler et al., 2002).

Sports and games are a way of maintaining physical fitness, apart from promoting recreation and showing one's upper-handiness over the other. Excelling the tactics of sports events, many legends have emerged in which led to its evaluation and development of different events. Throwing some light on the reasons and circumstances would help us in better understanding the field of sports.

Struna (1977), rightly point out that the sport has emerged under certain social context and it use understanding which becomes important to appreciate the true spirit of sports, again finds that there are multiple contexts in which the evolution can be explained based on the post experiences. Such exploration led to the study of a range of sports topics such as Physical Education, dance and play.

1.2 Softball -

Softball is a baseball descendant game that requires speed, strength, and endurance. Softball is a bat-and-ball sport that entails skills such as throwing, fielding, pitching, catching, base-running and hitting (Craig et al., 1985). To develop in these areas, many factors play a significant role like skill, practice, level of physical activity, techniques, psychological traits, nutrition and other environmental factors. However, a particular body size and shape and certain genetically conditioned abilities and features are required in order to achieve top level performance in softball. The increased competitiveness and the drive to excel in sports and to achieve national or international status in sports demands that players accomplish their optimum performance levels in a systematic way with clearly defined performance objectives and training programs. This has motivated coaches, managers, sports administrators, trainers, sports scientists and players to spend more time and resources assessing their sports performance more scientifically. This requires the assessment of anthropometric characteristics, physiological variables, and sports specific skills for adequate sport performance. Since performance in sports is a multivariate phenomenon, it may be influenced by physique in addition to physiological function, biochemical constraints, psychological state, environment and socio-cultural factors.

Softball changed into invented in 1887 in Chicago as an indoor recreation. It changed into at numerous times known as indoor baseball, mush ball, playground, soft bund ball, kitten ball, and, because women also played it, the sport became referred to as ladies' baseball. The name softball was given to the sport in 1926. There are two types of softball. In the most common kind, slow-pitch softball, the ball, now and again larger than the same old 12 inches, have to arch on its direction to the batter, there are 10 gamers in a crew, and bunting and stealing are prohibited. In rapid pitch softball, the pitch is speedy, there are 9 players on the sphere at one time, and bunting and stealing are accepted. Softball guidelines range quite from those of baseball.

Softball is a physically demanding sport comprised of several specializations such as hitting, fielding, throwing, pitching, base running etc. requiring diverse skills and different types of fitness. It requires upper extremity power, hand grip strength, excellent eye-to-hand coordination and the coordinated movements of the hips, shoulders, arms and wrists. It requires speed, strength and endurance (Terbizan et al., 1996). Softball players need a considerable amount of upper body muscle balance because of the specificity of underarm pitching and overhead throwing activities. Coaches and athletes consider that the forearm plays an important part in hitting, pitching and throwing the ball.

In softball, there is little scientific information to suggest which specific components of strength, speed or power best predict softball playing ability. There is also very scant information regarding the anthropometric characteristics as the predictors of performance in softball. As such, the development of performance prediction models as a function of specific tests of physical capacity and anthropometric characteristics may provide insight into those factors which relate to softball playing ability.

However, a particular body size and shape and certain genetically conditioned abilities and features are required in order to achieve top level performance in softball. The increased competitiveness and the drive to excel in sports and to achieve national or international status in sports demands that players accomplish their optimum performance levels in a systematic way with clearly defined performance objectives and training programs. This has motivated coaches, managers, sports administrators, trainers, sports scientists and players to spend more time and resources assessing their sports performance more scientifically. This requires the assessment of anthropometric characteristics, physiological variables, and sports specific skills for adequate sport performance (Carter, 1985).

Softball is a demanding sport, Involving batting, sprinting, throwing and catching. To achieve any success as a player one needs a high stage of coordination, agility, power and versatility. Improving your flexibility can be performed via incorporating normal softball stretches into your softball schooling application. Softball is a group sports invented by way of George Hancock in 1897 on the Farragut Boat membership in Chicago. Later, out door model as the sport known as Kiter ball turned into advanced by using Hewes Rober, Today Softball is played all over the sector with the aid of actually thousands and thousands of human beings from all walks of life. The Softball crew includes 15 players each, 9 gamers participate in the game and 6 gamers are in reserve. The in shape includes five, 7 or 9 innings.

Playing softball every day, and sometimes twice a day, takes a physically and mentally fit athlete. The ability to accept failure since it is an inevitable part of the game, while being prepared to perform in the moment that demands excellence takes a special set of skills physically and mentally. From pitching to fielding, and hitting to base running, the softball player must take care of their body through a properly implemented strength and conditioning program that understands the value of recovery, nutrition, and sleep.

The game of softball originated as a kind of baseball for youngsters. Several popular sports had a significant impact on the development of softball, including boating, boxing, and football. There was also a rowing club, boxing glove, and football club involved in the original game of softball.

A Harvard supporter made effective use of a broom handle by snatching the batter's baseball glove. As a consequence, an indoor baseball game was developed. At Chicago's Farragut Boat Club that day, George Hancock, a Harvard and Yale graduate, knotted a boxing glove to create a soft, big ball. Because soft balls did not need gloves, players could play barehanded. He began the game by using chalk to draw a little baseball field on the yacht club deck. The final score was 41-40. Hancock rushed to get a large soft ball and a bat with a rubber tip after being told of the regulations. This very bright kid even drew permanent foul lines on the floor of the Boat Club.

Indoor baseball became popular in the Chicago area during that harsh winter. Baseball players have traditionally relied on weight training to maintain their fitness levels over the summer. A winter league was founded in 1889. When the weather warmed up, softball was relocated to an outdoor field (they were all too small to qualify as baseball grounds). The games used big and soft balls with a diameter of 10 to 20 inches, similar to the ones used indoors. Lewis Rober, a fire fighter, was instrumental in the game's development. Because his city's fire fighters were on route to the firehouse, he decided to keep them occupied in the meantime. The parking lot had previously been situated next to the fire station. Unlike baseball, which has a bigger field and a harder ball, softball features smaller fields and allows for more offensive and on-field movement. The seven-inning game duration restriction was established to enable games to be finished in less than an hour while yet offering sufficient competitiveness and activity. As a result, competitions between other firehouses began drawing audiences of up to 3,000 people. Pickup soccer was formally sanctioned by the Minneapolis Park Board in 1913, and it was played across the city's public parks and playgrounds.

Eventually, the game spread across the upper Midwest and into Canada. In 1897, Toronto became the first city outside the United States to establish a softball league. The game was known by a number of different names. If the same game was played in several locations, it was referred to as a separate game (common names included cabbage ball, mush ball, and pumpkin ball). Despite the nomenclature change, the bat and ball dimensions, as well as the exact rules, remained almost same. The sport drew men and women of various ages and backgrounds. In 1931, Kids and Kubs travelled the nation as a softball team comprised of young children and teenagers. They stood out from the crowd since they were all at least 75 years old and dressed up for the game. The new name for the game, softball, was revealed in 1926, when the director of the Denver YMCA suggested it. When the 1933 World's Fair opened in Chicago, a sports goods company and a journalist agreed to co-host a softball competition. Men's fast pitch, men's slow pitch (a.k.a. sixteen-inch softball), and women's softball were the three tournament divisions. Around 350,000 spectators flocked to the stadium, which served as a hub for the world's fair. This was one of the first instances of softball having a global effect. Numerous websites facilitated the growth of teams, leagues, and tournaments in the United States and around the globe. Five million individuals are anticipated to have played softball during the seven-year period from now to 2022. In the years after World War II, Canadian troops popularised softball in the Netherlands. Softball was introduced in Australia for the first time in 1939. Softball made its debut in the United Kingdom in 1962, during the filming of A Touch of Class in London. In 1983, the United Kingdom's first women's softball league was created.

As a result of the tournament's success, the Amateur Softball Association was founded in the fall of 1933. The organisation established game-playing regulations, thus ensuring uniformity and fairness. Due to the fact that softball is accessible to individuals of all ages, a league established regulations for various age groups. The usual ball in many young leagues is a 12-inch variant. The age of the player and the pace of the game both have an effect on the size of the field. While the distance between the pitcher's mound and home plate varied between 35 and 50 feet, the distance between bases varied between 55 and 65 feet. International interest has been sparked in the fast pitch game. The Women's Fast pitch Softball League players throw the ball as quickly as big-league pitchers.

Three professional men's softball leagues were formed between 1977 and 1982, the first of which was the American Professional Slow Pitch League, which started play in 1977.

The International Softball Federation has governed softball play since 1951. In 1965, the world's first women's fast pitch softball competition was organised in Melbourne, Australia. The five-team event was won by the host team. The US women's team has won six of the eight competitions. At the conclusion of the year, Mexico City hosted the first men's world championships. The United States men's team won the title in 1966. Softball world championships have been contested every four years since 1970.

In 1996, women's softball was promoted to the Olympic level. Softball's technical development reached a new level during the 1996 Olympics, when the IOC provided money for cutting-edge biomechanical research on pitching throughout the games. The whole United States women's softball team has been successful in all three gold medal games. The popularity of women's fast pitch in the United States is reflected in the country's power in the sport. In the United States alone, Softball is played by around 25 million people. As a result of the International Olympic Committee's most recent meeting in Singapore in July 2005, softball and baseball were eliminated as Olympic sports for the 2012 Summer Olympic Games.

1.2.1 History and Development Softball -

Ball games of one type or another have played for thousands of years. However, the game of softball is relatively young ball game if one considers being "born" within the last one hundred years to be young. A descendent of baseball and an outgrowth of a combination of games, softball had its start at the Farragut Boat Club in Chicago, Illinois, in 1887 where players used boxing glove for a ball and a broom for a bat to play the game of "indoor baseball". George W Hancock, a recognized leader in the development of the new game, devised rules and provided a ball larger and softer than a baseball and a bat with relatively small head. The indoor game became very popular because it could be played in a rather small area, was inexpensive, offered a competitive experience. Initially the game was played in gymnasiums of colleges and Universities,

Young Men's Christian Associations (YMCA), and the Turnvereins, but the players soon moved outside the playing grounds. Each community devised its own rules and variations of the game were played under such names as Diamond Ball, Indoor Ball, Kitten Ball, Mush Ball, Playground Ball, Pumpkin Ball and Recreation Ball. Softball is fun to play; otherwise, the millions of participants would choose to young age in a different game. The two games, fast pitch and slow pitch, are challenging and thrilling to play and exciting to watch. Basically, they are very much alike; the major differences, as the names imply, are found in the velocity and trajectory of the pitch. The ball is always pitched under handed, but it maybe lobbed across the plate in a high arch or go passed the batter traveling eighty miles an hour. In both games offensive strategies are cantered on getting base and scoring runs while defence manoeuvres to prevent this (Jones and Murray, 1978).

Softball has had the most rapid and remarkable growth of all sports in the United States. Indoor baseball was the first played in the United States about 50 years ago, but the game was not played outdoors to any extent until the late

1920's, when the Canadians began to use it on small playground areas. The game developed rapidly in Canada and returned to the United States, where it received impetus from the National Recreation Association. This organization supported the game because of its similarity to baseball and its wide appeal to all ages. The game has been called playground ball, kitten ball, recreational baseball, ladies baseball and soft baseball; but in 1933, the name "Softball" was generally adopted and a national tournament was held at the World's Fair in Chicago. Leo H. Fischer and M. J. Pauley were largely responsible for forming the Amateur Softball Association, an organization that made a great contribution to the game by effecting standardization of the rules and conducting national tournaments which attracted teams from all parts of the country.

A softball field is small enough to be accommodated on most playgrounds. The bases are not as far apart as in regulation baseball, and the outfield boundary is much shorter. Because of the large ball (12 inches in circumference) and small bat, it is not possible to drive the ball as far as a hard baseball; hence diamonds may be located on areas that would be far too small for baseball. The large ball is less likely to cause finger injuries than the baseball, so the game while retaining many of the desirable qualities of regulation baseball, is particularly valuable for use on playground and recreational areas. Equipment generally is less expensive than that required for baseball, making softball more acceptable for use in informal groups often found in parks, playgrounds, churches, and other community groups. Although softball lends itself to such informal use, it can be and is played on a level requiring a high degree of individual skill and team strategy; where teams develop these qualities, the game has often provided an attraction for fans that rivals that of baseball (Seaton et al., 1969).

Softball is a physically demanding sport comprised of several specializations such as hitting, fielding, throwing, pitching, base running etc requiring diverse skills and different types of fitness. It requires upper extremity power, hand grip strength, excellent eye-to-hand coordination and therefore the coordinated movements of the hips, shoulders, arms and wrists. It requires speed, strength and endurance. Softball players need a substantial amount of upper body muscle balance due to the specificity of underarm pitching and overhead throwing activities. Coaches and softball players consider that the forearm plays a crucial part in hitting, pitching and throwing the ball (Singh, K. M. 2017).

1.2.2 History of Indian Softball -

Softball an active sport is played in many countries which is also an Olympic sport. It is played in many popular countries like Latin America (Cuba), Europe, China and Japan. One can witness that the active sport-softball is played even in India. It was played in Rajasthan for the first time. Dr. Darhrath Mal Metha, the father of Indian softball, formed the softball association of India on November 21, 1961. In America the international softball federation was formed in 1964. The first world softball championship (women) was organized by the international softball federation in Australia in 1965. The team visited India after the federation to promote softball game in different states; the first rule book on softball game was published in India in 1965. The first national softball championship for men and women was organized in the year 1967 at Jodhpur. The softball association of India was recognized in 1973 by All India Council of Sports.

A softball sport requires brisk and explosive use of the softball player's upper body, specially their arms strength. Softball player should rely just as heavily on their gym and personal trainer as they do their softball equipment, and should be in shape when softball season comes. Playing softball actually will improve your upper body strength and often will provide a "chiselled" look to the player's upper body. Holding the softball bat, swinging the softball bat, and finally making contact with the softball bat will challenge the entirety of the arms, especially forearms. Strength is being too derived from the softball player's core muscles in order to begin the swing and put the necessary power behind it. There are many methods of strengthening a softball player's core, including the use of medicine balls and free weights that are available at the gym. To work on speed and acceleration for running the softball bases, practicing sprints or "shuttle runs" is a magnificent method. Thus, physical fitness is an inseparable part of sports performance and achievements. The quality of its utilization value is directly proportional to the level of performance. That means the greater the level of fitness, the greater the ability of a person to attain higher levels of performance.

According to Maud and Foster (1995), anthropometrics is the science that deals with the measurement of size; weight, and proportions of the human body, as body size and proportions, physique, and body composition are important factors in physical performance and fitness. A standard anthropometrical analysis during a physical fitness assessment would involve determining the height, mass, somatotype and body fat percentage of an individual or athlete. Understanding the characteristics of anthropometric measurements in every field is a vital and powerful aspect in the performance of players. The knowledge of physiological personality might provide the player with information as to where training may be directed to compensate for areas where he/she may be below average in his/her specific sport (Callan et al., 2000; Zabukovec and Tiidus, 1995). The anthropometric measurements in sports provide opportunities to learn more about the responses of individuals to the demands of exercise and the anthropometrical adaptations to training which underlie improvements in an individual performance.

The performance in softball depends on interaction of individuals in tactical moves, the capability of players in the basic skills of hitting, fielding, throwing, base running and catching and in the more specific skills related with particular playing positions. In such contexts, the assessment of player performance must consider the physical attributes, as well as the tactical and technical aspects of performance.

1.2.3 Playing Method -

Softball is played between two teams on a large field, with 9 players from one team on the field at a time. The field is usually composed of a dirt or brick-dust infield that contains the quadrilateral shape and running areas, and a grass outfield. However, the field can consist of other solid and dry surfaces such as artificial turf or asphalt. There are 4 bases on the infield (first base, second base, third base, and home plate); the bases are arranged in a square and are typically 45 to 65 feet (13,7 to 19,8 meters) apart. Near the center of this square is the pitcher's circle, and within the circle is the "rubber", a small flat rectangular piece of rubber about a foot and a half in length. The rubber can be 40 or 43 feet away from home plate, depending on age level and the league one is playing in. The object of the game is to score more runs (points) than the other team by batting (hitting) a ball into play and running around the bases, touching each one in succession. The ball is a sphere of light material, covered with leather or synthetic material. It is 10 to 12 inches (or rarely, 16 inches) (28 to 30.5 centimetres) in circumference. The game is officiated by one or more neutral umpires. Players and umpires are generally free to ask for a brief stoppage at any time when the ball is not in play, or immediately following a play once its outcome is clear (Austen, Jacob, 2005).

The game is played in usually 7 innings. Each inning is divided into a top half, in which the visiting team bats and tries to score runs, while the home team occupies the field and tries to record 3 outs; then a bottom half, when the teams' roles are reversed. Youth leagues may play 6-inning games; most ASA and middle school rules specify 5 innings and a maximum time of 1.5 hours.

To start play, the offense sends a batter to home plate. The batting order must be fixed at the start of the game, and players may not bat out of turn. The defence's pitcher stands atop the rubber and throws "(pitches"), the ball towards home plate using an underhanded motion. If the pitcher is playing in a fast pitch league, they must have both feet in contact with the pitching rubber at all times. If the pitcher is playing in a slow pitch league, the pitcher is allowed to take one step back prior to releasing the ball during the forward movement.

The batter attempts to hit the pitched ball with a bat, a long, round, smooth stick made of wood, metal or composite. A pitch must fly over the home plate within an area known as the strike zone: it must cross over home plate, and as it crosses it must be above the knees and slightly below the shoulders (roughly the armpit or the shirt logo). The strike zone therefore varies from batter to batter, and the umpire behind home plate is the sole arbiter of what is or is not a valid pitch. A pitch outside the strike zone is a ball, and if the batter reaches 4 balls, the batter is awarded the first base. A pitch that crosses the strike zone is a strike, and a batter who reaches 3 strikes is out (a strikeout), and the next batter in the order comes to bat. A strike is also recorded on any pitch that the batter swings at and misses entirely, and also on a pitch that is hit foul (out of play). A foul ball may or may not result in a strikeout dependent upon what association and local league rules. However, bunting a foul ball does result in a strikeout. In some associations and leagues, bunting is not allowed and results in an out.

The batter attempts to swing the bat and hit the ball fair (into the field of play). After a successful hit the batter becomes a base runner (or runner) and must run to first base. The defence attempts to field the ball and may throw the ball freely between players, so one player can field the ball while another moves to a position to put out the runner. The defence can tag the runner, by touching the runner with the ball while the runner is not on a base. The defence can also touch first base while in possession of the ball; in this case it is sufficient to beat the batter to first base and an actual tag of the batter is unnecessary. A runner is said that to be "thrown out" when the play involves two or more defensive players. Runners generally cannot be put out when touching a base, but only one runner may occupy a base, at any time and runners may not pass each other.

When a ball is batted into play, runners generally must attempt to advance, if there are no open bases behind them; for example, a runner on first base must run to second base if the batter puts the ball in play. In such a situation, the defence can throw to the base that the lead runner is attempting to take (a force out), and the defence can then also throw to the previous base. This can result in a multiple-out play: a double play is two outs, while a triple play, a very rare occurrence, is three outs. Runners with an open base behind them are not forced to advance and do so at their own risk; the defence must tag such runners directly to put them out rather than tagging the base. A ball hit in the air and caught before hitting the ground, in fair or foul territory, puts the batter out. A fly ball is a ball hit high and deep, a pop fly is a ball hit high but short, and a line drive is a ball hit close to the horizontal. After the catch, runners must return to their original bases; if the defence throws the ball to that base before the runner returns, the runner is out as well, resulting in a double play. A runner who remains on the base until the ball is touched, or returns to the base (tags up) after the catch, may try to advance to the next base, at the risk of being tagged out between bases. As in baseball, the infield fly rule applies in some game situations to prevent the defence from recording multiple force-outs by deliberately dropping an easy catch.

Offensive strategy is mostly just to hit the ball skilfully to let the batter reach base and advance other runners around the bases to score runs. The count of balls and strikes indicates how aggressive the batter should be. The offense may try to sacrifice, with the batter deliberately making an out in order to advance runners. Defensive strategy is more complex, as particular situations (number of outs and positions of base-runners) and particular batters call for different positioning of fielders and different tactical decisions. The defence may decide to allow a run if it can achieve one or multiple outs.

1.2.4 Positions of Players -

There are nine players out on the field at one time. Although the pitcher and catcher have the ball the most, each person has a specific job. In the infield there is the pitcher, catcher, first baseman, second baseman, shortstop, and third baseman. In the outfield there is left fielder, center fielder, and right fielder. In slow pitch softball there is an extra outfielder in the outfield, who is specified as a roamer. Normally, the defensive team will play with four outfielders, meaning there is a left fielder, left-center fielder, right-center fielder, and right fielder.

Pitcher -

The pitcher is the individual who throws the ball from the middle of the diamond. As in baseball, fast pitch softball has a rubber used to control the pitchers delivery. A pitcher must have both feet on the rubber at the start of their delivery, but neither foot is on the rubber when the ball is released. In baseball the rubber is on a mound of dirt as gravity is used to generate speed. In softball, the pitchers throw from within a circle and the rubber in the circle is not elevated. The circle is also used to control base runners. When the pitcher is in possession of the ball in the circle, a runner who has passed a base must either advance or return to the base. They cannot “bait” the pitcher to throw or they will be called out. The pitcher usually throws the ball in the strike zone.

In softball, the pitcher uses an underarm motion to pitch the ball towards the strike zone. As soon as the pitcher makes a throw, the fielders are ready to field balls that are hit in the middle of the diamond. Pitchers usually tend to be tall, very flexible and have good upper body strength. Pitchers can be right handed or left-handed. The softball pitcher makes a windmill motion while throwing under hand, unlike baseball pitchers who throw over hand.

Catcher -

The catcher is normally behind home plate in a squatted position (some plays may require the catcher to stand at an angle for intentional walks). At the plate, the catcher is responsible for catching pitches, keeping mispitched balls in front of the plate, calling pitches that are normally done through hand signals, and they are considered the leaders of the field. Catchers must know how many outs there are, the number of strikes and balls on the batter so they can relay that to their team-mates. They must also know how many runners are on base and where the ball should be thrown next in the following play. Catchers are strong, need to be smart, quick on their feet, to accurate throws so they can pick runners off at each base.

First baseman -

The first baseman is the position to the right side of the diamond when facing the field. The major role of the first baseman is to receive throws from other defensive players in order to get a force play at first base. Another role, they play is to make fielding plays on all balls hit towards first base. The first baseman is usually involved in every play that occurs on the field. Individuals at first base have quick hands, a good reach and are always on the lookout to catch the player off base. They are also generally taller and left-handed throwing, which gives them an extended reach. First basemen however, can be both left and right handed. Just like the catcher, first basemen may wear mitts.

Second baseman -

The second baseman plays in between the first baseman and the gap at second. If the ball is hit to the left side of the field, second covers second base. If the ball is hit on their side of the field, they back up the fielder, cover first, or field the ball depending on where it is hit. The second baseman also is the cut off on balls hit to the right side of the outfield. The second baseman also throws directly to the pitcher to prevent runners from scoring. Mostly, when the ball is hit to the left side of the field, the second basemen plays second for shortstop. Also, in the case of a bunt, the second baseman must cover first as the first baseman runs to get the ball from the bunt, then it is often thrown back to first base.

Shortstop -

The short stop fields all balls hit to the infield between the second and third bases. This individual also helps cover second base, third base and is frequently involved in force plays, double plays and frequently throws the ball to the catcher to throw out runners at home plate. On steals to second base (when the runner from first is advancing to second on the pitch), the shortstop usually covers. The shortstop only doesn't cover second base when a right handed batter is up. In this case, the second baseman covers the steal. Most short stops are very quick, agile and think fast. They are the captains of the infield as is the center fielder of the outfield. Short stops may play in a restricted zone but are faced with many types of hits and interact closely with the 2nd base, 3rd base and home plate. Often double plays are due to quick thinking/reaction by the short stop. When a ball is hit up the middle and the shortstop catches it, they will flip the ball to the second baseman for the best result. Short stop also takes the cut off for the left field when the play is at second base. When the ball is hit to the right side of the outfield, the short stop then covers second base. If there is a runner on first base and there is a hit down the line or in the gap to right field, the shortstop will then go for the cut to third. Meaning the runner that is going from first to second will most likely be advancing to third, and in that case, the shortstop will be the cut off for the throw from right field to third base. Shortstop is one of the most difficult positions to play due to the amount of balls being hit in that direction (Andrea Blackshaw (2004).

Third Baseman -

The third baseman covers all hits to the third base. The third base person also frequently throws the ball to the catcher trying to throw out the runner at home plate. The individual at third base usually thinks fast because of the types of hits that commonly occur in that zone. Quick thinking can often stop runners from scoring or can result in double plays.

The out-fielders are players that cover the grassy area behind the infield. Outfielders are named for their positions in the field relative to home plate. Traditional outfield positions include a left fielder, a center fielder, and a right fielder. These players usually back up the plays made by the infielders and make plays when the ball is hit past or over the infield. Because their positions cover a greater range than those of infielders, outfielders tend to have strong throwing arms. Outfielders are responsible for throwing fielded balls to a player known as a cut-off person (usually a second baseman or short stop). In this way, the outfielders more efficiently return the ball to the infield with the goal of preventing base runners from advancing to additional bases. In some leagues / organizations, four outfield players are utilized by each team, with the center field position being shared between two players known as the left-centre fielder and the right-centre fielder (or, alternatively, the deep center fielder and the short fielder).

1.2.5 Rules of Softball -

Rule 1-

Sec. 1. Altered Bat -

A bat is constructed of wooden materials; however, metal handles are inserted into the bat to substitute for the wooden parts. The tape is applied excessive two layers thick on the bat, one to hold it together and one to identify it. Replacing the grip is not considered modifying the bat even if another lawful grip is used. An altered bat is connected to a bat with a flare or cone grip.

Sec. 2. Appeal Play -

In the event of a live or dead ball appeal, the umpire cannot reach a decision unless the opposing teams do not debate or raise an objection. When a coach, manager, or captain enters the playing field, only a dead ball appeal is permitted.

Sec. 3. Ball Compression -

ASTM standard test procedure for evaluating Softball compression-displacement, called Standard Test Method for Compression-Displacement of Softballs, determines load in kilograms (pounds) needed to compress a softball 6.5mm (0.25 inches) as measured.

Sec. 4. Ball COR -

Measured using the ASTM test technique for determining the Coefficient of Restitution of softballs, the A softball's restitution coefficient.

Sec. 5. Base Line -

"A straight line is the basic line."

Sec. 6. Base on Balls -

"A base on balls is scored when the plate umpire determines that four pitches, including illegal pitches, are balls. The hitter is allowed to take a free base. This is referred to as a complimentary item."

Sec. 7. Base Path -

A base route is a straight path that links a base to the runner's present position when a defensive player is about to try (or has recently tried) to tag the runner.

Sec. 8. Batted Ball -

When not intending to hit the ball, any batted ball is in play.

Sec. 9. Batter -

A batter is an offensive player whose main goal is to help his team in scoring runs while he is in the batter's box. He remains a hitter until he is struck out or becomes a batter runner.

Sec. 10. Batter's Box -

The batter's box is the position of the batter where the hitter is physically confined in order to assist his team score runs. Also known as being inside the batter's box, the lines are in baseball, this is referred to as being within the batter's box.

Sec. 11. Batter-Runner -

While still in the batter's box, a batter-runner is a player who has not yet been struck out or reached first base while the game is still in progress.

Sec. 12. Batting Order -

This request should be fulfilled, since it includes details on when the players will be awoken to take their positions in the batting order, as well as all the positions needed to be filled. In order for the players to participate, they must enter the batting order.

Sec. 13. Bunt -

While trying to avoid the hit, the batter missed the ball but touched the bat and was eased through the infield.

Sec. 14. Catch -

There is just one term for an outfield catch when the fielder catches the ball in his or her hand or glove: a legal catch.

Sec. 15. Catcher's Box -

The catcher must wait until the pitch is released before he leaves his position. If the pitch hits the ground, it must be considered a live ball. In the catcher's box, the lines are regarded to be the boundaries.

Sec. 16. Chopped Ball -

When the batter hits the ball to the ground in the cuts, the ball bounces up in the air.

The host team chooses whether team will be visitors or home teams by mutual agreement through the coin. When a batter strikes the ball to the ground in the cuts, it bounces up into the air.

Sec. 17. Coach -

It is the entire duty of the team to represent the other team and the arbitrary team at the moment of the infield. If the coach is not there at the time of the match, the player may be coached.

Sec. 19. Ball Dead -

A dead ball is one that collides with an object that is not part of the official equipment or playground, or with a player/person who does not play, or who plays in the arbiter's or inappropriate clothing, or who has proclaimed the umpire dead.

Sec. 20. Team Defensive -

The defensive team is the field team.

Sec. 21. Ball Delayed Dead -

A delayed Dead Ball is a gaming situation in which the ball remains alive until the game's end. When the game is completed, the arbiter must signal a dead ball and enforce the applicable rule. (See Section 3 of Rule 9)

Sec. 22. Player Designated (DP) (FP Only) -

The Designated Player is a beginning offensive player who fights for the FLEX PLAYER on the line-up card at tenth (10th) place.

Sec. 23. Base Dislodged -

A dislodged base is one that has been displaced from its original position.

Sec. 24. Play double -

A double play is a defensive game in which the continuous action legalises two attacking players.

Sec. 25. Ejection from the Game -

Ejection is any arbitrator's act of expelling a player, an official, or a team member from the game and serves as a cause to break the rules for the remainder of the game.

Sec. 26. Fair Territory -

Playing area from the house plate to the bottom of the field fence, including the first base and the third base foul line.

Fair area is that portion of the playground from the home base upwards to the bottom of the extreme playground clamps, includes the foul lines at first and third base.

Sec. 27. Fielder -

Any defensive team member on the field is a fielder.

Sec. 28. Fly Ball -

A hitter struck the ball in the air, called a fly ball.

Sec. 29. Player Flex -

The Flex Player is the starting pitcher to whom the Designated Player (DP) pitches and whose name appears in tenth position on the line-up card.

The name of the designated player appears 10th on the line-up card. DP comes just for batting.

Sec. 30. Out Force -

At that time the batter runner reached 1st base that the first base runner is forced to decide because the runner should leave base before the batter reaches base. Batter hit the ball but runner of first does not go forward to 2nd base at that time. It's an appealing game.

Sec. 31. Forfeit -

A forfeit is the act of the plate arbiter which says that the game is finished by proclaiming the winner of the non-offending team.

Sec. 32. Foul Ball -

A hit ball that is technically the foul –

- lies between a home plate and 1st base plate or between home plate and 3rd base.
- Bound beyond or over foul area 1stor 3rdon.
- In 1stor 3rd base first touches in foul area.
- If the ball hits the equipment, the arbiter's clothes, or any player item or anything alien to the natural soil. The time ball in the foul area is called a foul ball.
- If the ball hits the batter's hand or the batter's hand (s)

Sec. 33. Home Team -

The host team chooses whether team will be visitors or home teams by mutual agreement through the coin.

Sec. 34. Illegal Bat -

The bat that is not produced by the illegally bat regulation.

Sec. 35. Pitcher Illegal -

An Illegal Pitcher is the legal player in the game but cannot tune the pitch as if the manager removes the pitching position or the umpire because the defensive conference limit has been surpassed.

Sec. 36. Illegal Player -

An Illegal Player is the player that has either a defence or an offence on the board card that was not disclosed to the arbiter. These are the kinds of illegals:

- A Replacement Player that enters the game (under substitute player regulation) without reporting to the umpire plate.
- A Withdrawn Player (returns to the match without reporting to the umpire plate (per the replacement player regulation).

Sec. 37. Batted Illegally -

When the hitter strikes the ball fair or foul, an illegal ball happens:

Sec. 38. Caught Illegally Ball -

When a fielder's hat, mask, glove, or any other part of his uniform gets caught in a hit ball, thrown ball, or pitched ball when it is removed from its appropriate position, it is considered an unlawfully caught ball.

Sec. 39. Player Ineligible -

An ineligible player is one who is no longer legally engaged after being dismissed by the umpire. An unlicensed player is no longer a player, although he or she may continue to play as a coach.

Sec. 40. Ineligible Player Replacement -

A player who is not authorized to replace a player who must leave the game to treat a bleeding injury is not permitted to participate in the game. There are two types of ineligible replacement players. 1. Has been dismissed or expelled from the game by the arbiter due to a violation of the rules. 2. Is already in the line-up.

Sec. 41. In Flight -

In flight, this includes any batted, thrown, or pitched balloon that has not yet landed on the playground, as well as any other item other than the fielder.

Sec. 42. Infield -

The infield is the section of the playground that encompasses areas usually covered by infielders.

Sec. 43. Infield Fly -

A body of knowledge A fly is defined as a fair fly ball that can be picked up by a normal infielder if the first and second bases, or the first and second and third bases, are occupied before two players go out. For the purposes of this regulation, the pitcher, catcher, and any outfielders who enter the play field are considered infielders.

Sec. 44. Inning -

An inning is described as the part of a game in which the teams blend attack and defence. Each team has three outs in each inning. A new inning starts as soon as the third player leaves the previous inning.

Sec. 45. Interference -

1) An offensive player or team player who disrupts, obstructs or confuses a defensive player by trying to perform a play.

2) An arbitrator who hinders the effort of a catcher to toss a runner off the base.

3) A spectator who enters the ground obstructing a fielder to play the ball or who contacts the ball on which an infield player attempts to play.

Sec. 46. Legal Touch -

A lawful touch (tag) is a fielder action to touch (or tag):

Sec. 47. Ball Legally Caught -

A caught ball occurs if a hit, thrown or pitched player is caught, but only because he is not caught any portion of his uniform like cap, helmet, etc. in the field. It must be taken and held firmly in the hands or in the glove.

Sec. 48. Drive Line -

A line drive is the ball in the air that is directly and forcefully hit into the playground.

Sec. 49. And Line-Up Card -

A line-up is a collection of players who are presently playing offensive and defensive roles in the game, including DPs, flex players, and EPs (SP ONLY). The Line-up Card contains the starting players' last name, first name, position, and uniform number, as well as the team manager's last name, uniform number, and first and last names.

Sec. 50. Obstruction -

Obstruction is defined as the act of a defensive player or team member preventing a batter from hitting or smashing a ball.

Any fielder who prevents a runner or batter from progressing while running legally by not holding the ball or not playing a batted ball, making a fake tag without the ball or having the ball and pushing the runner off the base, or holding the ball but not acting on the runner, prevents the runner from progressing while running.

Sec. 51. Offensive Team -

The team at bat is the offensive team.

Sec. 52. Batter On-Deck -

The batter on the deck is the offensive player whose name is followed by the batter's name.

Sec. 53. On Deck Circle -

The circle on the deck is closest to the player's bench, where the batter may wake up or practise before entering the batter's box.

Sec. 54. Three Meter Line -

The last half-distance between home plate and the first level where a batter runner must operate is three feet (one metre). to avoid being penalized for interfering with a thrown ball from the plate area or trying to receive a thrown ball while on the first level of the building

Sec. 55. Option Play -

An optional game is a game in which the offensive coach has the ability to take illegally the opposing team or play outcome. These are the following choices.

1) Obstruction of the catcher ii. Use of an unrecognized glove iv. Illegal replacement
2) Unlawful pitch v. An illegal pitcher came back to play and to throw. A batter woman on a two-out walk.

Sec. 56. Outfield -

The outfield is the portion of the playfield outside the base lines or region usually not covered by the inferior diamond shaped. It also covers 3rd bases and borders of the playfield in foul lines beyond 1 stand.

Sec. 57. Slide over -

An over-slide is an offensive pad when he slides over a base he is attempting to reach as a runner. It always happens when his velocity leads him to detach from the base, which then puts him at risk. The battery runner may pass through the 1st base without any risk. The only requirement is that he should promptly return to that base.

Sec. 58. Overthrow -

A reversal is a game where a ball is tossed from one player to the next. It means that the boundary lines of the ground are crossed or ii. obstructed.

Sec. 59. Ball Passed (FP Only) -

A passed ball is a phenomenon when pitch is retained or managed with simple efforts by the caught person.

Sec. 60. Pitch -

A pitch is the action of the pitcher delivering the ball to the batter.

Sec. 61. Circle Pitcher (Only FP) -

Circle of the pitcher 2.44 m from the middle of the plate. The lines are included in the circle.

Sec. 62. Foot Pivot -

a) The pivotal foot (FP ONLY) is the basis on which the pitcher pulls the plate away from him.

b) (SP ONLY) which, when the lifting plate is put in touch with the lifting plate, must be in contact with the pitching plate until the lifting ball is released.

Sec. 63. Ball Play -

Usually, the phrase "play ball" arbitrarily resumes the action.

Sec. 64. Community -

A protest is the activity of a defensive or offensive team that objects, either by an umpire, to interpreting or implementing the rule of the game, or by the eligibility of a member of the team roster.

Sec. 65. Player Replacement -

A player who has to leave the game to take part in an injury that causes bleeding is obliged to join the game for a specified amount of time.

1) The replacement player may be a new player who has not yet entered the game, a returning player who entered the game but was later replaced by the game, or a beginning player who has left the game and is no longer eligible for re-entry.

2) A substitute player shall not be summoned as a substitute provided that the substitution is notified to the arbitrator.

Sec. 66. Runner -

Runner is a player, that is, a hitter with a spin on bat and a free base on the first base. The opposing team doesn't take him out.

Sec. 67. Starting Players -

The starting line-up consists of players identified on the official line-up given to the Chief of the Umpire and/or the plate umpire.

Sec. 68. Stealing -

Stealing is the action of a runner who tries to move on to the hitter during or after a pitch. Stealing in Slow Pitch is not permitted.

Sec. 69. Strike Zone -

The strike zone is this zone between the armpits of the batter and the top of his knees as he stands.

Sec. 70. Substitute -

A replacement is a player on the official line-up slip on the final list who is not in the squad when it resumes.

Sec. 71. Tagging Up -

Tagging up implies that an infield player touches a runner returning to his base.

Sec. 72. Team Member -

A member of the team comprises a player, physiotherapist, coach, team manager, doctor, etc. They may sit on the team bench.

Sec. 73. Throw -

The action by the fielder is a throw away when it throws the ball towards another fielder.

Sec. 74. Time -

When the ball is dead, the umpire orders the play to be suspended by using the word "time."

Sec. 75. Trapped Ball -

A trapped ball implies a lawfully battered fly ball that reaches the playground or the net before the adversary is captured.

Sec. 76. Triple Play -

There are three offensive players in a continuous defensive action game called as triple game.

Sec. 77. Wild Pitch (FP Only) -

Wild pitch is such a high, so low or so broad pitch that cannot be stopped or controlled by the catcher. (FP ONLY)

Sec. 78. Wild Throw -

When an infield field tosses a ball in such a way that it cannot be readily captured or stopped by his team-mate, such a throw is termed a wild throw.

Article 79. Sec. Withdrawn Player -

A player withdrawn indicates that the player is removed from the game and is recorded as a line-up according to the regulation.

Rule 2 - Field of play -

Sec. 1. The Playing Field -

1) Is the region where the ball may be played and played legally?
2) There is an open space outside of the stupid lines between the home plate and the backstop.

The Double Base Rules -

1) The ball hit in the field is termed fair play, while the ball struck the field is called foul play.
2) When the batter misses the third strike to hit the ball, if the opposing team calls before the batter rider returns to fair position
3) A defensive player must always utilise just the fair share of the base.

4) The battery runner must return to the real section after the base has been executed.

5) When the hitter reaches both bases, he has to sprint 2 bases by touching the base part.

6) The tagging of a fly ball must be utilised for the proper part.

7) When attempting to pick-off, the runner must get back to the right part.

8) If the batter goes to the right part it should stand on the basis if it is determined that there is no touch with the player's base or location the runner is known to be out.

9) Tagged with the ball

10) The runner leads of the foul portion of a pitched ball.

Rule 3. Equipment -

Sec. 1. The Official Bat -

a. Should be manufactured in a single piece, be permanently assembled in several pieces, or be composed of two or more replaceable components. If the bat is made of replaceable parts, it must satisfy the following requirements:

b. In order to avoid unrecognized material mating components, a unique locking key must be provided.

c. All material combinations must satisfy the same requirements as a. One or more bat pieces or a part of a one-piece bat if removed is merged.

d. The grain direction of each component should be identical to the slugger, if one, two or more pieces of wood consisting of a timber glued to adhesive (bat).

e. Made of plastic, bamboo, carbon, graphite, magnesium, glass fibre, pottery.

f. Can be covered but should only include wood or glue and have a transparent finish.

g. It ought to be spherical and smooth. f. The length must not exceed 86,4cm, nor shall the weight exceed 1077,0 g (38 ounces).

h. Shall not exceed 5.7cm diameter (2 1⁄4 inches) in its biggest section. A tolerance of 0.80 mm (1/32 in) is possible.

i. Metal should be cornered.

j. If pins, rivets, rough or sharp angles or other external attachment that poses a danger are not revealed. A bat of metal should be devoid of burrs and fractures.

k. It is used for handling metal.

l. Should have a cork grip covered by tape (no smooth, plastic tape) or the composition substance. A safety grip must be at least 25.4cm (10in) long and no longer than 38.1cm (15in) from the bat's small end. Resin, pine tar, or spray applied to the safety grip to enhance grip are only permitted on the grip.

m. The bat shall include a safety button with a minimum length of 0.6 cm (1/4) projecting from the handle of the bat at a 90-degree angle. N. The maker should permanently append the weight distribution and bat length when it is manufactured. The producer shall not change the dimensions above at any expense thereafter.

Sec. 2. Game Ball -

a. The ball should have a regular, smooth, hidden stitch with a seam.

b. Should have top quality cover No. 1 tanned chrome horse hide.

c. Softballs used in ISF should comply with ISF specifications.

Sec. 3. Gloves & Mitts -

a. All infield players are using a glove except for the catcher.

b. The glove should not be lacquered or webbed. Glove length is 12.7 cm.

c. The colour of the glove should be different from the colour of the ball.

Sec. 4. Player Shoes -

For all participants, the shoes should be obligatory. Shoes may be made out of leather or canvas

Sec. 5. Protective -

a. Masks-Mask is mandatory for caught

b. Face Masks - The team members may wear the authorized face mask of plastic.

c. Body Protectors - The body protection caterer should be used to prevent injury.

d. Shin Guards - The player stands behind the batter to prevent knee damage to the patella and shin bone shin guards. Helmets - For batters, helmets are important.

Sec. 6. Playing Field Equipment -

The unofficial equipment in the playing arena is not allowed.

Sec 7. Uniform -

Every player should wear the same uniform type. Coach may have various uniform colours.

a. The players should utilise CAPS-Caps. b. UNDERSHIRTS - Undershirts must be of similar variety.

b. Trousers - Softball game is permitted with long or short pants.

c. Numerals - On the back of the shirts, contrasting colour numbers should be printed at least 15.2 cm high. 01 to 99 numbers may be used.

d. Names - Player names may be put on the back of the T-shirt.

Rule 4. Coaches, players and replacements -

Sec. 1. Trainers -

You should sign the line-up card by the head coach. 2. ROSTERS/LINE-UP CARDS. A. Batting line-up cards to the score table should be provided. c. Separate rosters for male and female athletes are to be utilised.

Sec. 2. Players -

In terms of position A softball team should consist of the following elements: 1. There were nine players involved: the pitcher (F1), the catcher (F2), the first baseman (F3), the second baseman (F4), the third baseman (F5), the shortstop (F6), the left fielder (F7), the centre fielder (F8), and the right fielder (F9) (F9). Nominee for Player (DP). Ten players: the same as fast pitch but with the addition of one designated player.

Sec. 3. Players Starting -

A starting player is deemed official by permission of the plate arbiter if they are on the line-up card.

Sec. 4. Designated Player -

Also termed DP is a DESIGNATED PLAYER. He comes to bat at any moment.

Sec. 5. Extra Player -

An EXTRA PLAYER is known as DP, too. He should be registered in the first line-up.

Sec. 6. Illegal Player/Substitutes -

A substitution can be done at any stage of the game, provided he/she shall be mentioned in the starting line –up.

Sec. 7. Disputed Calls -

Every participant on the pitch should not object to the arbitrary choices. In such scenario, the arbitrator warns the player. The recurrence may lead to the player being expelled.

Sec. 8. Conduct Dugout -

Coaches, players, replacements or other bench members should not leave the assigned seat.

Rule 5 – The Game -

Sec. 1. Turn At Bat Choice -

It is chosen to spin coin hitting or fielding.

Sec. 2. Ground Fitness -

The suitability of the ground for a game is determined only by the umpire plate.

Sec. 3. Game Regulation -

a) A regulation game consists of seven innings.

b) A tie regulation game is declared if the score is equal at the end of five or more completed inputs.

c) These regulations do not apply to player or spectator acts that may necessitate the game's termination. If an umpire is physically attacked by a member of the team or a fan, the umpire may lose the game.

d) A tied game is repeated from the start if the initial line-up is altered.

Sec. 4. Game Winner -

In a regulatory game, the side that pays the most wins. In the event of incomplete entries, the winner is determined based on the results of the final entry.

Sec. 5. Run Ahead Rule -

a) All tournaments and championships must employ a leading rule.

b) (Only SP) Twenty (20) run after five (5) innings after four (4) innings or fifteen (15).

c) Unless the bat team second obtains the required number of runs, the whole innings should be played. When the team that bats first is in the top half of the inning, the team that bats second must be allowed to bat in the bottom half.

Sec. 6. Tie-Breaker -

In case of tie-breaker, a not out hitter of the previous inning advance to the 1st base, the next batter will come to bat.

Sec. 7. Scoring of Runs -

When a runner touches all of the bases and home planes before the final start, he or she earns one run for their efforts.

Rule 6. Regulations on pitching -

Sec. 1. Preliminaries -

a) Before the delivery (pitch) begins, the pitcher may not pitch on or near the pitcher's plate if he or she does not have the ball in his or her possession.

b) Should not be taken into account in the pitch unless the catcher can receive the pitch.

c) At the moment of the pitch, both feet should contact the pitch plate.

Sec. 2. The Pitch Start -

The pitch continues after one hand has been removed from the ball.

Sec. 3. Legal Delivery -

a) The pitcher must not move to throw until the ball is delivered to the hitter immediately.

b) While starting the pitch, the pitcher swings the both hands.

c) The pitcher must not utilise a windup in which the forward motion is stopped or reversed.

d) The pitcher should not take two complete pitch strokes.

e) The pitch should be underneath the arm.

f) The pitcher may take one stride forward and the back foot must be ground contact.

g) After the arbitrary call, the pitcher has to throw in 20 seconds.

Sec. 4. The Catcher -

At the time of pitch the catcher should be in the catcher box and toss the ball back into a pitcher after every pitch save for the strikeout or runner.

Sec. 5. Base Throwing -

After the pitch position comes, the pitcher should not throw to the base. When this is done, the appeal is invalid.

Sec. 6. Intentional Base on Balls -

If the team wishes to provide a hitter with free base, it may be done by informing the umpire plate.

Sec. 7. Pitches Warm-Up -

The teams are given up to 5 spots before the first inning or change of the pitcher and 3 spots thereafter as warm up spots to the catcher or any other team member.

Sec. 8. No Pitch -

If a player, manager or coach appeals for time before a pitch pits during the ban or before the batter takes his or her position or early base appeal, then no pitch and ball is dead shall be proclaimed.

Sec. 9. Ball Dropped -

When the ball is pitched out, the game goes on and a ball is proclaimed.

Sec. 10. Illegal Pitcher -

Unlawful Pitcher cannot return for the rest of the game.

Rule 7. Flapping -

Sec. 1. The Batter On-Deck -

The hitter should be near his team bench in the deck circle until the umpire calls. He can warm up with up to 2 bats.

Sec. 2. Position Batting -

After the umpires call for the ball, within 10 seconds, the hitter should assume his place in the batter's box.

Sec. 3. Strike Called by the Umpire -

a) After a legal pitch, the hitter passes past the strike area without hitting the ground.

b) After the ball, the batter struck and missed.

c) After every tip of foulness. Ball stays in play, but if it's third strike, the hitters are out.

d) If the hitter has fewer than two strikes after each faulty ball.

e) After each pitched ball, which contacts any portion of the batter struck and missed.

f) After a ball has been lit and the batter has been struck when the ball seems to be in the strike zone on the field.

g) After an umpire calls for a ball and a hitter fail within 10 seconds to reach the batter's box.

h) If the member of the batting team intentionally erases the batter's box lines.

Sec. 4. A Ball is called up by the Umpire -

a) When the ball was thrown off the strike zone, or reached the earth before it reached the home plate and was not swung or touched the home plate and the hitter did not swing. The ball continues to be played and runners who are responsible may advance.

b) If a ball strikes the batter beyond the strike zone

c) If the warm up pitches are more than the limit, then every excessive warm-up pitches. Dead call ball and runners don't go on. If the ball is not returned by the pitcher immediately

Sec. 5. The Batter is Out -

a) If the third strike is swinging or missing or the ball contacts a portion of the batter, or if it doesn't swing, the ball strikes the hitter during the strike area.

b) If it is found that a batter is using a modified or unlawful bat, enter the batter box. If the foot of the batter is fully beyond the box lines of the batter and the contacting ground or the house plate is outside a foot when the ball is struck.

c) If the batter bunts foul after the second strike, the batter will be called out.

d) If you call a third strike. If the hitter is bolting or chopping down the ball.

e) If the ball is struck on fair territory a second time with the bat.

f) If the catcher caught a third strike that was called or swinging.

g) If there are fewer than two batters with three strikes and the first base is occupied.

Rule 8 –

Sec. 6. Runner is out -

a) Upon being lawfully touched by the ball while off the base, or being thrown off the base before reaching first base after the third strike is dropped, the batter runner is ruled to be out of the ballgame.

b) If a fly ball is caught before it reaches the ground, or if it is caught by any other object or person other than a fielder.

c) The hitter gets tagged off the base or thrown out after a fair ball before it reaches the first base if the batter has not moved on to the first base. It's hit after a fair ball.

d) After the hitter hits the ball, if he simply touches on his first base attempt a reasonable part of a double base and a base playing is created.

e) When the hitter went back to home plate to evade or delay a fielder tag.

f) If he merely touches the fair part of the base while utilizing double bases in a force play scenario and strikes a jug Who is ready to grab a thrown ball and who is also making use of the fair portion of the base in order to get points?

g) If someone enters the field who is not a member of a team and interferes with the fielder who is about to catch or is able to catch a fly ball in the arbiter's judgement.

Rule No. 9 Sec. 1: Dead Ball - Ball in Play -

Sec. 1. The Ball is Dead -

a) The ball is no longer alive. Rule number nine.

b) In the following situations the ball is dead and not at stake.

c) If the ball batted unlawfully.

d) If the pitcher picks up the signal and the hitter move from one box to another

e) If the pitch has been ruled no pitch, a thrown ball is touched if part of the batter's gear is dropped.

f) If the batting team causes an interference.

g) The umpire or runner gets hit before to contact pitcher or fielder by a batted ball.

h) If the ball travelled outside the playground.

i) The replaced player may continue to the given bases. The replaced player shall touch bases not previously affected.

j) Whether the thrown ball is being bunted or battered.

k) If a backstop pitch or passed ball.

l) While in the box of the batter, if any portion of the batter hits with its own ball, After every ball or strike.

m) When Time is called and a defensive player appeal.

n) If the hitter doesn't enter the batter's box 10 seconds after the umpire shouts PLAY BALL.

Sec. 2. The Ball is in Play -

In the following situations the ball is at stake.

a) When the pitcher has the ball in his pitching position and the plate umpire signals "PLAY BALL," the game is said to have begun. Every inning and a half, every inning and a half.

b) When the Infield Fly Rule is in effect, a fly ball must be caught in the infield in order to be considered caught.

c) Whenever a thrown ball travels through and stays inside the playing area of a field.

d) Whenever a fair ball hits an arbiter or runner on fair ground, the game is declared a forfeit.

e) Except for the pitcher, no other pitcher could make a pitcher or touch a field, including a loudspeaker.

f) When a fair ball strikes a ground arbiter.

g) As soon as the runners reach the bases to which they are entitled, an illegal ball is hit, thrown, or spun away from the field.

h) When a rider is summoned for a previous rider.

i) When a blocked runner is not played, the ball remains live through till completion of the play.

j) If a fair ball is hit lawfully.

k) If a runner has to go back in reverse order, while the ball is playing.

l) When a runner has the right to a base before being removed.

m) h. When a foundation is removed, the runners proceed around the foundations.

n) If a runner travels more than 0.91m in regular or reverse order from his base route, avoid getting touched by the ball in a field's hand.

o) If a cyclist gets tagged or forced off the bike.

p) When the arbitrator calls out the runner, he or she will not return to the base when play restarts after a pause.

q) When a valid live ball appeal is filed.

r) If the batsman makes contact with the ball.

s) When a live ball is hit, a photographer, a groundskeeper, or a police officer is sent to the field of play.

t) When a fly ball is lawfully caught and returned to the pitcher.

u) When an attacking player comes into touch with the ball after it has been thrown.

v) If the hitter lowers his or her bat and the ball rolls into the fair area against the bat, and the umpire's decision was not made with the purpose of tampering with the ball, the batter or the batter's team may challenge the decision.

w) When a thrown ball makes contact with an undetermined object.

x) Unless the ball is declared dead in accordance with Section 1 of this rule.

y) When a thrown ball makes an accidental contact with a coach.

z) Once a batter is called for a ball and four balls are called, the batter may not be removed before reaching first base.

aa) When the batter was summoned and three strikes were called on the batter.

Rule 10. Umpires -

Sec. 1. Power and Duties -

Arbitrators may and must apply each provision of these rules in their role as Members of the Organization or Association to which they have been nominated have been appointed. In order to enforce one or more of these rules and apply the penalties stated in them, they have the power to instruct a player, coach, captain, or management to take whatever action they believe is required. In any event that is not covered by the regulations, the umpire plate has the authority to make decisions.

Sec. 2. The Plate Umpire -

Will be behind the catcher. He will be responsible for and accountable for the proper conduct of the game.

Sec. 3. The Base Umpire -

a. In accordance with the proper arbitration procedures, certain places on the field should be assumed as required.

b. Will assist the plate to enforce the rules of the game in every way feasible.

Sec. 4. The Responsibilities of a Single Umpire are as follows -

When just one arbitrator is chosen, his duties and power must cover all aspects. The umpire starts from behind the house plate for each pitch. The umpire must go off of the platform onto the field in order to have the best position for each hit ball or action that takes place.

Sec. 5. Umpires Change -

If an arbitrator is not incapacitated by illness or disease, the opposing teams cannot agree to switch umpires in a game.

Rule 11. Campaigns -

Sec. 1. Complaints that will not be entertained -

The acceptance or consideration of protests must be denied if they are mainly based on the accuracy of an umpire's decision, or if the team that filed the protest won the game.

Sec. 2. Protests That Will Be Received -

Protests received and evaluated include the following kinds of issues

a) An erroneous understanding of the rules of a game.

b) Failure of an arbitrator to apply the correct rule to a particular instance.

c) Failure to apply the appropriate punishment for a specific offence.

d) Team roster members eligibility.

Rule 12. Scoring -

Sec. 1. The Official Scorer -

a) Ensure compliance of each game in accordance with the procedures outlined in the following regulations.

b) The sole entity having the ability to make all judgement decisions should be the Supreme Court. When determining whether a batter's advancement to first base is the result of a hit or an error, for example, it is the scorer's duty. In contrast, the scorer is not permitted to make any decisions that are inconsistent with the Official Playing Rules or the judgement of an umpire.

1.2.6 Effectiveness -

Stretching has been found both effective and ineffective based on its application for treatment.

Although many people engage in stretching before or after exercise, the medical evidence has shown this has no meaningful benefit in preventing specifically muscle soreness. It may reduce the lactic acid build up in the muscles, making the next workout more bearable.

Stretching does not appear to reduce the risk of injury during exercises, except perhaps a dynamic warm-up for runners. While running places extreme stress loads on the joints, static stretching can help to improve joint flexibility. However, this has not been proven to reduce risk of injury in the runners. A dynamic (stretching) warm up has been shown to help overall running performance.

Delayed-onset muscle soreness, also known as DOMS, typically arises 48 hours after an exercise bout. Stretching before or after the exercise did not show any significant benefits in the onset of DOMS.

There are different positives and negatives for the two main types of stretching: **static and dynamic**. Static stretching is better at creating a more intense stretch because it is able to isolate a muscle group better. But this intensity of stretching may hinder one's athletic performance because the muscle is being overstretched while held in this position and, once the tension is released, the muscle will tend to tighten up and may actually become weaker than it was previously. It has been shown in high level athletes, such as gymnasts, after performing a static stretching routine that it has negative effect. The gymnasts lost the ability to jump vertically as high as prior as well as no improvement in their straddle jump or flexibility. Also, the longer the duration of static stretching, the more exhausted the muscle becomes. This type of stretching has been shown to have negative results on athletic performance within the categories of power and speed. However, to be able to do usual daily activities, a certain amount of range of motion is needed from each muscle. For example, the calf muscles are one of the muscle groups that have the most need for adequate flexibility since they are deeply related to normal lower limb function. When the goal is to increase flexibility, the most commonly used technique is stretching. Chronic static stretching was shown to increase range of motion of Dorsiflexion or bringing one's foot closer to their shin by an average of 5.17 degrees in healthy individuals versus 3.77 degrees when solely using ballistic stretching. While static stretching is shown to decrease power and speed in higher level athletes, when it comes to the older population who live more sedentary lifestyles static stretching has been shown to increase muscles strength and power.

Still, ballistic stretching is likely to increase flexibility through a neurological mechanism. The stretched muscle is moved passively to the end range by an external force or agonist muscle: holding a muscle in this position might reduce muscle spindle sensitivity, with repeated stretch applied at the end range inhibiting the Golgi tendon organ.

Dynamic stretching, because it is movement-based, may not isolate the muscle group as well or have as intense of a stretch, but it is better at increasing the circulation of blood flow throughout the body, which in turn increases the amount of oxygen able to be used for athletic performance. This type of stretching has shown better results on athletic performances of power and speed when compared to static stretching.

However, both of these types of stretching have been shown to have a positive impact on flexibility over time by increasing muscle and joint elasticity, thus increasing the depth and range of motion an athlete is able to reach. This is evident in the experiment "Acute effects of duration on sprint performance of adolescent football players". In this experiment, football players were put through different stretching durations of static and dynamic stretching to test their effects. They were tested on maximum sprinting ability and overall change in flexibility. Both static and dynamic stretching had a positive impact on flexibility but, whereas dynamic stretching had no impact on sprint times, static stretching had a negative result, worsening the time the participants were able to sprint the distance in.

1.3 Stretching -

Stretching is a physical exercise that requires putting a body part in a certain position that'll serve in the lengthening and elongation of the muscle or muscle group and thus enhance its flexibility and elasticity.

The stretching of a muscle fiber begins with the sarcomere, the basic unit of contraction in the muscle fiber. As the sarcomere contracts, the area of overlap between the thick and thin my filaments increases. As it stretches, this area of overlap decreases, allowing the muscle fiber to elongate. Once the muscle fiber is at its maximum resting length (all the sarcomeres are fully stretched), additional stretching places force on the surrounding connective tissue. As the tension increases, the collagen fibers in the connective tissue align themselves along the same line of force as the tension. Therefore when you stretch, the muscle fiber is pulled out to its full length sarcomere by sarcomere, and then the connective tissue takes up the remaining slack. When this occurs, it helps to realign any disorganized fibers in the direction of the tension. This realignment is what helps in the rehabilitation of scarred tissue.

The initial changes that are produced by stretch training involve mechanical adaptations that are followed by neural adaptations, which contrasts with the sequence observed during strength training.

1.3.1 Effects of Stretching -

- Improves flexibility delaying impaired mobility associated with aging
- Improves performance in physical activities as a result of improved flexibility
- Reduce the risk of injury and damage
- Improves blood circulation reducing muscle soreness and time for recovery
- Increasing the range of motion.

1.3.2 Types of Stretching -

Stretching exercises have traditionally been included as part of the training and recovery program. It's important to note that, maximal strength, number of repetitions and total volume is different for each type of stretching.

The three different types of stretching are:

- Static stretch (SS)
- Dynamic stretch (DS)

- Pre-contraction stretching: Proprioceptive Neuromuscular Facilitation stretching (PNFS).

1) Static Stretching -

Static stretching (SS) is a slow-paced controlled physical activity which involves putting the body part in a comfortable position that elongates the muscle without causing pain with low force for a prolonged duration of time (usually 30 seconds).

There are two types of static stretching -

1. Active-static stretching involves performing static stretches without assistance.
2. Passive-static stretching involves performing the stretches passively with assistance from an external force which can be a partner, an accessory or the force of gravity.

Effects of static stretching-

1. Improves range of motion in the joint
2. Reduces stiffness and pain in muscles
3. Reduces the risk of muscle strains and other injuries
4. Improves postural awareness and body posture
5. Increases circulation which in return decreases recovery period after exercise

Duration of Static Stretching -

The duration of the hold of the stretch is irrelevant to notice improvement but rather how many times the stretch is repeated in a week. It's important to note that each muscle should be stretched only once and should be held for five minutes which is broken into five one-minute exercises or ten exercises of thirty seconds. The more we stretch in a week, the better the outcomes where according to certain studies stretching for more than three weeks served in decreasing stiffness and increasing the range of motion (two to eight-minute stretches served in increasing range of motion while ten minute stretches served in re-obtaining normal range of motion).

For individuals whose main objective is general fitness, it's recommended that static stretching should be done at least twice a week and stretch held for a minimum of 15 seconds followed by dynamic stretching.

For older adults, the duration of static stretching should be longer to notice improvements. For example, a sixty-second hold stretch served in an increase by 2 degrees per week in range of motion while a thirty-second hold stretch served in an increase by 1 degree.

For children, incorporating static stretching in physical education class significantly improved the flexibility of the hamstring muscles where four sessions per week served in a 17 degree increase in range of motion while two sessions per week improved the range of motion by 9 degrees.

2) Dynamic Stretching -

Dynamic Stretching (DS) is a controlled movement, unlike Ballistic stretches that involve bouncing movement and increases the risk of injury, which involves the performance of a movement progressively increasing the range of motion through successive repetitive motions till the end of the range is achieved. It can be done standing or while moving.

For individuals who perform specific sports such as swimming, dynamic stretching exercises involve mimicking the movement of the activity such as circling arms before getting in the water and is often done after static stretching.

Effects of dynamic stretching -

- Restore physical functioning and flexibility

- Improve neuromuscular control through repetitive movement which: enhance the nervous message conduction speed, motor control and muscle compliance
- Elevates core body temperature
- Accelerates energy production
- Improves performance measures such as speed and strength.

Based on a randomized controlled trial conducted on 60 participants on the efficiency of dynamic stretching on hamstring extensibility and stretch tolerance compared to static stretching, dynamic stretching was more effective at achieving an increase in hamstring extensibility and stretch tolerance one-hour post-intervention than static stretching.

Muscles used in Softball -

The primary Muscle groups within the lower body used whilst playing softball are the quadriceps, hamstrings, hip flexors, groin and the calf muscle groups. The most important muscular tissues inside the top body consist of the rotator cuff muscle tissues of the shoulder and the middle muscle mass of the abdominals and lower returned.

1.3.3 The Benefits of Softball Stretches -

Coaches place a Lot of emphasis on together with softball stretches in the numerous softball training packages. This is due to the huge wide variety of advantages of stretching, which includes: Regular softball stretches are key to retaining flexibility, which in flip fosters an awesome posture and reduces decrease back ache and discomfort.

Stretching sports in softball schooling can increase a participant's athleticism via controlling muscle imbalances, which reason pulled muscle groups and also contribute to clumsiness, which in itself can cause damage.

Even the most effective softball stretches, as a part of your typical softball education application allows promoting blood movement, which improves intellectual alertness and coordination. Finally, regular softball stretches can assist save you injuries.

1.3.4 Stretching Routine for Softball Players -

No matter what position you play in softball, you'll help prevent injury and ensure your body is warmed up and ready to play by engaging in stretching exercises. Stretching exercises help increase blood flow to all parts of your body and it also increases flexibility and range of motion. Your coach might have specific exercises for you to perform before game time or you can do self-directed stretching exercises.

Rotator Cuff

Warm up your shoulder joint and rotator cuff muscles by doing a rotator cuff stretch before every game. This stretch warms up the entire shoulder and encourages flexibility and full range of motion. To do this stretch, stand with your feet shoulder-width apart. Place your right hand behind your back, with the top of your hand against your spine. Grasp your right elbow with your left hand and gently pull your right elbow forward. Hold the stretch for five seconds and then release. Switch sides and repeat.

Hip Rolls

Hip rolls help loosen the pelvic muscles, stretch the hip joints and increase flexibility in your lower spine. This exercise is done lying on the ground on your back. Bend your knees and stretch your arms out at shoulder height. Slowly drop your knees to the left side of your body, but keep your left hip firmly planted on the ground. Hold the stretch for several seconds and then lift your knees and rotate them to the right. Repeat this back-and-forth roll 10 times.

Dynamic Stretches

Stretches that encourage motion are performed through dynamic stretches. Such stretches are not static but rather encourage movement while warming up muscle groups. You can perform softball dynamic stretches with slow running butt kicks and high knee raises while slow jogging. To perform a butt kick, take small steps forward, lifting first your right leg and then your left behind you high enough for your heels to touch your buttocks. Perform a high knee jog while jogging forward, lifting your knees to the waistline, alternating knees while moving forward.

Shaking Arms

An effective loosening stretch for the hands, arms and shoulders can be accomplished with a shaking arms exercise. Have a partner stand in front of you, holding your hands. Gently yet firmly, your partner shakes your hands up and down. Keep your wrist, elbow and shoulder joints loose during the shaking motion. If you prefer, shake one arm at a time and then switch sides. Shake each arm for approximately 10 to 20 seconds for an effective warm up and stretch.

1.3.5 The Three Best Softball Stretches -

Stretching is one of the most beneath-utilized techniques for enhancing athletic overall performance, stopping sports activities injury and properly rehabilitating sprain and stress damage. Don't make the mistake of questioning that something as easy as stretching won't be effective.

Below are 3 of the excellent stretches for softball; obviously there are plenty more, however those are a first rate area to start. Please make unique note of the commands with each stretch, and if you presently have any persistent or habitual muscle or joint pain please take extra care when appearing the stretches under, or seek advice from your doctor or physical therapist before appearing any of the following stretches.

Lying Knee Roll-over Lower Back Stretch: While Lying on your returned, bend your knees and allow them to fall to 1 aspect. Keep your fingers out to the facet and permit your again and hips rotate with your knees.

The lying knee roll-over lower back stretch explores loosening up restrictions in both the lumbar spine (lower back) and thoracic spine (mid-back). An inability to rotate the spine effectively will impact your ability to throw and hit well in softball, so including the lying knee roll-over lower back stretch into your stretch routine is an effective way to promote flexibility.

Here are the steps:

Lay on your back on your Yoga Strong Mat and hug your right knee into your chest. Let your right arm extend wide out to the right.

Using your left hand, guide the right leg up and across the body to the left.

Now, let your knee fall to the left. Bottom leg stays straight. Release your left arm to the left and if comfortable, bring your gaze to the right.

Hold for 5 breaths. Repeat on the other side

Elbow-out Shoulder and Rotator Stretch: Stand with your hand in the back of the middle of your returned and your elbow mentioning. Reach over together with your different hand and gently pull your elbow ahead.

Rotating Forearm and Wrist Stretch: Place One arm directly out in the front and parallel to the floor. Rotate your wrist down and outwards and then use your different hand to further rotate your hand upwards.

1.3.6 Performance -

Performance sports aim at high sports performances and for that the physical and psychic capacities of sportsmen are developed to extreme limits. This normally does not happen in other areas of human activity. As a result, performance sports yield valuable knowledge about the limits to which human performance and various performance factors can be developed. It also leads to discovery of means and methods for improving various physical and psychic capacities (performance factors) to exceptionally high levels. This knowledge can be fruitfully applied to other areas of sports and human activity.

Excelling within the overall performance of his or her selected sport is the most important intention of any elite athlete. The force to win, the choice to succeed and the ambition to push past the present limits of overall performance are all vital features of accomplishing excellence in elite sport. Softball players must continuously strive to obtain height tiers of overall performance to reach and finally stay on the pinnacle. In discipline sports activities, players must now move faster, expect higher, demonstrate more levels of technical and tactical capacity and persist longer than competitors from the beyond. The commitments made via membership, teach and participant in attempting to gain perfection undeniably necessitate an intensive quantity of time and monetary contribution, particularly as the space among winning and dropping grows ever smaller. The foundations for schooling and competing can no longer be based on easy subjective perspectives of how nicely softball players carry out or on traditional methods passed from one technology of coach to any other.

Coaches and sports scientists generally use performance assessments to evaluate four major facets of player performance; these are physical, mental, tactical and technical skills. It is also worth mentioning that performance assessment has direct relevance in the clinical setting for the diagnosis and prognosis of conditions as well as being used to evaluate the effectiveness of medical or exercise interventions that are designed to be therapeutic.

Variations in overall performance may also occur because of physiological, physical, tactical or psychological reasons (Marshall and Elliott, 1999). For instance, physiological factors together with fatigue, power and strength can impact on the way in which players perform physical skills, as can psychological factors which include stress and self-confidence. Poor anticipation and decision-making capabilities can equally affect players to make the incorrect tactical decision or to pick out an inappropriate method for the situation offered. Finally, maturational factors may also significantly affect the way in which technical talents are performed.

1.3.7 Sports Performance -

Sports performance is indeed an aspect of complex human performance which has several aspects or dimension several disciplines of sports science are required to work in a coordinated manner to explore the nature of sports performance and the process of improving sports performance. In the last few decades several disciplines of sports science have been established e.g., sports medicine, sports physiology, sports training, sports bio-mechanics, sports psychology, sports pedagogy, sports nutrition and so on. The sports science is still in its state of infancy and in the near future more sports science disciplines are likely to emerge some of these disciplines have already emerged, e.g., sports statistics, sports bio-chemistry, sports neuro-physiology, sports cybernetics etc.

Sports performance is the unity of execution and result of a sports action or a complex sequence of actions measured or evaluated according to socially determined and agreed norms. Sports performances are achieved in training or in competition. Sports competition is a social phenomenon which has evolved historically. In a competition several persons compete with one another. In order to decide the ranking of these persons, some criterion has to be used. The selection of this criterion is a function of the society. Out of a number of available criteria society can select any one or more criteria to decide or judge the performance.

1.3.8 Softball Performance -

Our softball performance training is delivered in a small group setting with a maximum of 8 athletes per session. Depending on the time of year, our sessions range from 45-75 minutes in length.

At the beginning of each session, we provide detailed demonstrations of each exercise to prepare our athletes to perform every exercise with the proper technique and know where they should feel each movement.

From there, we lead a dynamic warm-up that improves athletes' posture, flexibility, and mobility, as well as warming up the muscles and nervous system for the training to come.

Next, each athlete will pair up in groups of 2-3 and begin the workout. Each workout is designed, at varying degrees, to build strength, develop power and speed, enhance coordination, and increase joint and tissue resiliency.

Finally, we cool down to down-regulate the athletes' nervous systems and have them leave calm and relaxed. This optimizes the athletes' post-workout recovery period so they're fresh and ready for what comes next, dominating on the field.

Off-Season -

During the off-season, little to no competitive softball is played. During this phase of training, gaining strength, developing speed and power, and putting on lean muscle mass become the top priorities. This is when massive transformations are made.

In-Season -

During the season, which is a long and truly physical and mental battle, the athletes' training shifts. In this phase of their development, athletes' training sessions are shorter, still intense to maintain their abilities, and are filled with joint and arm care drills to keep the body healthy.

Pre-Season -

In the pre-season, the aim of training is to apply all the strength and power that's been developed and translate it into more specific movements related to softball performance. This phase is vital for players to feel out and refine any changes to their technique.

Post-Season -

The post-season is a short timeframe where athletes' should be prioritizing recovery and dumping any built-up fatigue accumulated during the season. In this phase of training, athletes work on foundational movements and begin their physical and mental prep for their off-season training.

1.3.9 Importance of Softball -

Softball is a game wherein hundreds of thousands of members are engaged for fun. The two video games, fast pitch and gradual pitch, are difficult and exciting to play and thrilling to watch. Basically, they are very lots alike, the essential variations, as the name mean are located in the pace and trajectory because the pitch, the ball is constantly pitched underneath passed, but it is able to be lobbed throughout the of past due in a excessive arch are go speedy the better. The interesting action appeals to the Utah of the world, the competencies worried in the game are easy herbal and are incredibly stimulating and satisfying to anyone who participators in the game.

This study is intended to check scientifically, the affiliation of these parameters with the playing ability of softball gamers. The motor health variables velocity, agility, strength, cardio-vascular staying power, response time, shoulder power and stomach energy had been associated with the gambling ability of softball players in this take a look at. A softball recreation calls for surprising and explosive use of the softball player's higher frame, especially their fingers. While softball player must rely just as closely on their fitness centre and personal instructor as they do their softball equipment, and need to already be in shape while softball season comes.

Actually playing softball will improve your higher frame energy and regularly will offer a "chiselled" look to the participant's upper frame. Holding the softball bat, swinging the softball bat, and finally making contact with the softball with the softball bat will project the entirety of the arms, specifically forearms. Strength is be too derived from the softball player's core muscular tissues if you want to begin the swing and placed the vital energy behind it. There are many techniques of strengthening a softball participant's middle, together with the use of medicine balls and unfastened weights that are available at the gym. To work on speed and acceleration for strolling the softball bases, training sprints or "commute runs" is a top notch approach. With your instructor, find an open space either indoors or out.

1.4 Purpose of the Study -

The research purpose is a statement of comparative analysis of effectiveness of stretching on performance of the softball players, the study is being conducted, or the goal of the study. The goal of a study might be to identify or describe a concept comparative analysis of effectiveness of stretching on performance of the softball players predict a situation or solution to a situation that indicates the type of study to be conducted performance of the softball players.

The game of softball is no exception. Softball is a very versatile game that makes enormous physical fitness, psychological, technical and tactical demands, but it is the physical fitness component that will directly determine the level of demand that can be placed on the technical, tactical, physiological and skills of a player.

Of late playing sport has become more tactical than ever before. Apart from strength and power approaching opponent with strategies and counter strategies is found to be highly essential. Therefore training assumes a great importance. Similarly now-a-days, many scientific training methods are found to be influence softball players' performance, because team performance mainly depends upon individual performance. Softball need to achieve certain fitness components, to be better performance.

The purpose statement identifies the variables, population and setting for a study. Every study has an explicit or implicit purpose statement. The research purpose should be stated objectively or in a way that does not reflect particular biases or values of the researcher.

1.5 Statement of the Problem –

A Comparative Analysis of Effectiveness of Stretching on Performance of the Softball Players.

1.6 Definition of Key Terms –

1. Stretching -

Stretching is a physical exercise that requires putting a body part in a certain position that'll serve in the lengthening and elongation of the muscle or muscle group and thus enhance its flexibility and elasticity.

2. Performance -

Performance of an individual is the individual efficiency in executing the skills and techniques in a fruitful manner during game situation which will be advantageous both for him-self and to the team for which he is playing. Performance will be assessed on 10 point rating scale; the rating will be based on subjective evaluation to predict the performance ability of softball players.

3. Softball -

Softball is a variant of baseball played with a larger ball on a smaller field. The game of softball originated as a kind of baseball for youngsters. Softball is a variant of baseball played with a larger ball on a smaller field. It was invented in 1887 in Chicago, Illinois, United States as an indoor game. It was at various times called indoor baseball, mush ball, playground, softball, kitten ball, and because it was also played by women, ladies' baseball. The name softball was given to the game in 1926, because the ball used to be soft.

1.7 Objectives -

The objectives formulated in the present study were as follows-

1. To find out the significant difference of effectiveness of stretching on performance of the softball players
2. To find out the significant difference of effectiveness of stretching on performance of softball players of rural area.
3. To find out the significant difference of effectiveness of stretching on performance of softball players of urban area.

1.8 Hypotheses -

The hypotheses formulated in the present study were as follows-

1. There is no significant difference of effectiveness of stretching on performance of the softball players.
2. There is no significant difference of effectiveness of stretching on performance of softball players of rural area.
3. There is no significant difference of effectiveness of stretching on performance of softball players of urban area.
4. There is no significant difference of effectiveness of stretching on performance of boys' softball players.
5. There is no significant difference of effectiveness of stretching on performance of girls' softball players.

1.9 Delimitations -

The study was delimited in the following aspects-

1. Subjects of the present study were delimited to 170 softball players (Girls & Boys) at college level at Bikaner Region's rural and urban area.
2. Various stretching exercises are used for players.

3. The player's age ranged from 20 to 30 years.
4. The present study will be delimited to Softball players who had participated in Inter Collegiate and Zone tournaments.

1.10 Limitations -

The following factors are considered as limitations in the present study-

1. No special technique was used to motivate the subjects during the administration of the tests.
2. The basic of knowledge of the subjects in exercise science and their previous experience of physical activities were not taken into consideration.
3. Since the subjects were motivated verbally during testing and training periods no attempt was made to differentiate their level of motivation.
4. The heredity factors of the subjects and their influence on the selected variables were not taken into consideration.
5. Subject answers to statements in the questionnaire were influenced by a variety of variables, including the statements' level of knowledge, the seriousness and sincerity of the subjects, and so on.

CHAPTER-II
REVIEW OF RELATED LITERATURE

2.1 Introduction -

Study of the related literature comprises locating, reading and evaluating reports of research as well as reports of casual observation and opinion that are related to the individuals planned research report. A study of relevant literature is an essential step to get a full picture of what has been done with regard to the problem under study. The investigator has made an attempt to bring a brief review of research related to the present study to form the background for the present study.

A study of relevant literature is an essential step to get full picture of what has to be done with regard to this study. Such reviews bring about a deep insight and clear perceptive of the overall field. The review will help the development of research procedure. The term literature is employed to include anything appropriate to the topic, such as theories, letters, documents, historical records, government reports, newspaper account, empirical studies, and so forth (James, 1993).

Any research work needs to be validated with the strength drawn from various relevant studies, available literature and methodologies adopted to conduct the research. Accordingly chapter II presents the relevant literature reviewed in connection with the present study.

According to Merry Hanrahan (1998) learning level of cognitive segment effects the inter-related factors such as control of the teacher and beliefs of the students he conducted a research on 11th class of biology and found that both intrinsic and extrinsic motivation could lead to deep involvement in learning are constrained by a preponderance of teacher centered methods of instruction. He concluded that a personal perspective is the non-linearity of the development of the theory.

The researcher finds out some of the review of literature which could be very supportive and strengthen this study. After going through the available literature, the investigator has presented some of the observations and findings of experts in this area.

The essential aspect of a research is the review of the related literature. In the words of Good, "The key to the vast store house of published literature may open the doors to sources of significant problems and explanatory hypothesis, and provide helpful orientation for definition of the problem, background for selection of procedure, and comparative data for interpretation of results". In order to be truly creative and original, one must read extensively and critically as stimulus thinking.

For any research project to occupy a place in the development of a discipline, the researcher must be thoroughly familiar with both previous theory and research. The literature related to any problems helps the scholar to discover what is already known, and would enable the investigator to have a deep insight, clear prospective and a better understanding of a chosen problem, and various factors connected to the study. So a number of books, journals, and websites were referred. In the following pages, an attempt has been made to present briefly a few of the important researches and studies conducted abroad and in India, as they have significant bearing on the present study.

The literature in any field forms the foundation upon which all future work will be built. If we fail to build upon the foundation of knowledge provided by the review of literature, the researcher might miss some works already done on the same fields.

2.2 Studies on related study -

Lloyd Smith (2008) The performance of baseball and softball bats has progressed markedly over the last 4 a long time. This has stimulated many associations to broaden test methods and measures to modify bat performance. The present have a look at reviews the development of laboratory bat overall performance tests. The check involves an to begin with desk bound bat this is allowed to balk after being impacted by using a ball. Bat performance changed into shown to be insensitive to changes in the manner the bat was supported and sensitive to the rebound ball and bat pace. The rebound ball and bat speeds had been in flip stimulated by air turbulence and bat vibration, respectively. Normalizing bat performance for the effect of ball weight, on the other hand, decreased its effect on bat overall performance.

Sayers et.al., (2008) conducted a study to determine which phase of a 30-m sprint (acceleration and/or maximal velocity) was affected by prepare performance static stretching. Data were collected from 20 elite female soccer players. On two non consecutive days, participants were randomly assigned to either the stretch or no-stretch condition. On the first day, the athletes in the no-stretch condition completed a standard warm-up protocol and then performed three 30-m sprints, with a 2-minute rest between each sprint. The athletes in the stretch condition performed the standard warm-up protocol, completed a stretching routine of the hamstrings, quadriceps, and calf muscles, and then immediately performed three 30-m sprints, also with a 2-minute rest between each sprint. On the second day, the groups were reversed, and identical procedures were followed. One-way repeated-measures analyses of variance revealed a statistically significant difference in acceleration ($p < 0.0167$), maximal-velocity sprint time ($p < 0.0167$), and overall sprint time ($p < 0.0167$) between the stretch and no-stretch conditions. Static stretching before sprinting resulted in slower times in all three performance variables. These findings provide evidence that static stretching exerts a negative effect on sprint performance and should not be included as part of the preparation routine for physical activity that requires sprinting.

Zakas et.al., (2008) conducted a study to examine in field conditions the acute effects of passive stretching after a general warming-up bout as well as the effects of passive stretching alone. Eighteen adolescent team soccer players participated in this study performing 3 different flexibility-training protocols in separate training sessions. In the first treatment stretching protocol a general warm-up was performed where subjects jogged for 20 minutes. The second treatment stretching protocol consisted of the same general warm-up followed by passive stretching of the lower extremities and the trunk, whereas the third and final treatment stretching protocol consisted of passive stretching alone, without any jogging. Passive range of joint motion was examined in hip flexion, hip extension, hip abduction, ankle dorsiflexion, knee flexion and trunk flexion using a goniometer and a flexometer.The general warming-up session induced a significantly increased range of motion only at the ankle dorsiflexion joint ($P<0.05$). Results also suggest that passive stretching alone and passive stretching after a general warming-up bout both induced a significantly increased range in all lower extremity joints and trunk flexion ($P<0.001$).

Sally j. Ford (2009) An intervention method changed into evolved, carried out, and evaluated that aimed at minimizing overall performance anxiety. The goal changed into to guide NCAA Division I softball athletes in the use of a respiration method that, by contributing to the management of overall performance anxiety, might help every athlete attain full potential on the softball field. The strategy cantered on the effects of the respiration technique at the individuals' heart fees, with regards to each day tension events; a coronary heart rate display and tension logs have been used to reap facts. All 4 of the athletes studied indicated improvement at numerous stages within the program. Some athletes upward push to the project, at the same time as others end up aggravating and may fail. It isn't always unusual for a instruct to witness an athlete succumb to such anxiety.

Yanagisawa et al. (2009) conducted a study to determine the anthropometric characteristics of Japanese baseball and softball players. 56 professional baseball players (26 pitchers and 30 position players) and 39 elite amateur female softball players (9 pitchers and 30 position players) were studied. All the subjects were assessed for height, weight, upper and lower limb lengths and circumferences and percent body fat. The players were also measured for subcutaneous adipose and muscle thickness of the limbs and abdomen and muscle cross-sectional area of the thigh and trunk with the help of a B-mode ultrasound apparatus and a magnetic resonance imaging system, respectively. Baseball pitchers showed significantly greater values than position players for body fat percentage, thigh and waist girths, triceps skinfold thickness, thigh skinfold thickness and calf skinfold thickness and muscle thickness of the subscapular and anterior site of the thigh. Softball pitchers were found to have significantly greater body weight, upper and lower limb lengths and circumferences, subscapular, thigh and abdomen skinfold thicknesses, and muscle thickness of the forearm, anterior site of the upper arm and anterior site of the leg than the position players. The baseball players showed significantly greater cross-sectional area in non-throwing side of the trunk muscle than in throwing side, particularly in lateral abdominal muscles. Softball players reported more differences between pitchers and position players in anthropometric measurements as compared to baseball players. On the other hand, baseball pitchers exhibited side-to-side differences in trunk muscle cross-sectional area compared to softball pitchers.

Dickson et al. (2009) assessed physical (height, weight, body composition, lean body mass, right and left hand grip strength) and performance (batted-ball velocity, throwing velocity, running speed, and vertical power) characteristics of the female softball players. 19 female softball players from the National College Athletic Association Division (NCAA) II and 16 female softball players from NCAA Division III were studied. The results of the study revealed that the NCAA Division II and NCAA Division III softball players did not differ from each other in physical and performance characteristics except for the batted-ball velocity suggesting that the NCAA Division II softball players were able to hit the ball harder than NCAA Division III softball players.

Lowe et al. (2010) studied the relationship of various physiological and anthropometric characteristics with the bat swing velocity of the college softball players. 20 female NCAA Division-I softball players were tested for height, weight, lean body mass, percent body fat, fat mass, grip strength and vertical jump. The bat velocity of the players was measured with the help of a SETPRO SPRT5ATM chronograph. The results revealed that there were significant moderate positive relationship between bat velocity and body mass, lean body mass and body fat. It is concluded that significant relationship exist between body composition and bat velocity.

Nimphius (2010) examined the effect of a two week cessation from resistance training on force, velocity and power in elite softball players. 7 elite female softball players were measured for counter movement jump. The results showed that there was a significant increase in velocity at peak power occurred from 96 hours to 14 days post training and a moderate to large effect reported in peak velocity, relative peak force, relative peak power and velocity at peak power. The correlation between jump height and peak velocity increased from 0.741 to 0.908. The relationship between jump height and velocity at peak power increased from 0.743 to 0.873.

Cahill and Jones (2010) assessed the relationship between body composition and performance tests in female volleyball and softball players. 16 volleyball and 18 softball NCAA Division-I female players were assessed for body composition, vertical jump, bench press, front squat, hang clean, pro agility, t-test and shuttle run. The results showed that in volleyball players, a significant positive correlation was reported between shuttle run times and body composition. A significant negative correlation was observed between body composition and vertical jump in softball players. There was a significant positive correlation between body composition and pro agility run times and shuttle run times in the softball players. Hang clean, bench press and front squat were not significantly related to body composition in both volleyball and softball players.

Teichler (2010) studied the relationship between bat velocity, upper and lower extremity power and the rotational kinetic chain in female softball players. 31

female players from NCAA Division II softball teams were measured for height, weight, arm length, medicine ball hitter's throw, seated medicine ball shot put, vertical jump and maximal bat velocity. Maximal velocity of the bat's sweet spot was measured using a 7-camera motion analysis system. The results showed that seated medicine ball shot put was the significant predictor of linear bat velocity and accounts for 6.6% of the variance in bat velocity. Body weight was also significantly associated with the linear bat velocity among the softball players.

Sophia Nimphius Et Al (2010) The purpose of this study Became to analyze (a) the cross-sectional relationship of electricity, power, and performance variables in educated female athletes and (b) determine if the relationship between these variables adjustments over the course of a season. In end, BW and relative power have strong to very sturdy correlations with pace and change of route capability, and these correlations stay consistent over the course of the season. However, it seems as though many relationships vary with time, and their relationships have to therefore be investigated longitudinally to better decide if those move-sectional relationships clearly mirror a deterministic relationship.

Nimphius et al. (2010) investigated the cross-sectional relationship of strength, power, and performance variables in female softball players and determine if the relationship between these variables changes after a season. 10 female softball players from a Australian Institute of Sport were examined for maximal lower body strength (one repetition maximum [1RM]), peak force, peak velocity and peak power during jump squats unloaded and loaded, unloaded countermovement vertical jump height. 1st base and 2nd base sprint performance and change of direction performance on dominant and non-dominant sides were also measured. The testing sessions occurred pre, mid, and post a 20-week training period. Pearson product-moment correlation coefficient was used to explore the relationship among different variables. The results showed that there were significant relationships of performance measures across all time points i.e. pre, mid and post a 20-week training period with body weight, speed and change of direction measures and relative strength and measures of speed and change of direction ability among the softball players. The vertical jump height did not significantly correlated with any measure of performance at any time point. It is concluded that the weight and relative strength of female softball players have very strong relationship with speed and change of direction ability across all time points i.e. pre, mid and post a 20 week training period.

Fletcher and Colombo (2010) conducted a study to explore the potential mechanisms underlying performance changes linked to different warm-up stretch modalities. Twenty-one male collegiate-semi professional soccer players (age, 20.8 +/- 2.3 years) performed under 3 different warm-up conditions: a no-stretch warm-up (WU), a warm-up including static passive stretches (SPS), and a warm-up incorporating static dynamic stretches (SDS). Countermovement jump, drop jump, peak torque, heart rate, core temperature, movement kinematics, and electromyography (EMG) were recorded for each intervention. Significant increases ($p < 0.001$) in performance were recorded for the countermovement, drop jump, and peak torque measures when the SDS was compared with the WU and SPS trials. When mechanism data were analysed, heart rate was significantly higher ($p < 0.001$) in the SDS condition compared with the SPS and WU conditions (a pattern also shown with core temperature), whereas the WU condition heart rate was also significantly higher than the SPS condition heart rate. When EMG data were examined for the rectus femoris muscle, significantly greater ($p < 0.01$) muscle activity was observed in the SDS condition compared with the SPS condition. It seems the most likely mechanisms to explain the increase in performance in the SDS condition compared with the SPS condition are increased heart rate, greater muscle activity, and increased peak torque.

Nimphius et al., (2010) conducted a study on the relationship between Speed, Strength, power and Change of Directional Performance of Female Softball

Players. Investigation of (a) cross-sectional relationship of performance variables and strength, power in trained female athletes and (b) find out if the relationship between these variables changes over the course of a season was the main aim of the study. The testing sessions occurred pre, mid, and post a 20-week training period. Ten female softball players (age = 18.1 6 1.6 years, height = 166.5 6 8.9 cm, and weight = 72.4 6 10.8 kg) from a state Australian Institute of Sport softball team were tested for peak force (PF), peak velocity (PV), maximal lower body strength(one repetition maximum [1RM]), and peak power (PP) during jump squats unloaded and loaded, unloaded countermovement vertical jump height (VJH)1 base and 2 base sprint performance and change of direction performance on dominant and non dominant sides. Pearson product–moment correlation coefficient at each testing session was used to establish the Relationship between body weight (BW), relative strength(1RM/BW), VJH, relative PP, relative PF, PV, speed, and change of direction variables. There were no significant relationships between VJH and any measure of performance at any time point. However, Substantial relationships were found across all time points with BW, speed, and change of direction measures (r = 0.70–0.93) and relative strength and measures of speed and change of direction ability (r = 20.7320.85). It can be concluded that relative strength and BW have strong correlations with speed and change of direction ability, and these correlations remain consistent over the course of the season.

Koley & Kumaar, (2011) had carried out current investigation to evaluate the selected the handgrip strength in Indian university softball players as dependent on selected hand-anthropometric variable assets. This cross-sectional study was based on 243 randomly selected Indian inter-university softball players As per the study, there were 121 males and 122 females and also they included with the untrained controls which were about 98 males and 102 females aged between 18-25 years. The following assessments were recorded as follows: Body height and mass, right hand width and lengths of 2nd, 3rd and 4th digits and right handgrips strength. The study shows that there was a difference of

$p<0.01$-0.001 in all the parameters except body height, right hand length and right 2nd digit length among all these four sets of data. Comparatively both the male and female softball players the partition of R2 into the contribution of the study of individual analysis and revealed the body height 18%, hand width would be around 11% and body mass of 8% and hand length of 5% which totally gave the specific data for the significant study. This data might play a very important role in the selecting the unique and best talented identification for softball.

Till et al. (2011) conducted a study to determine the relationship of performance variables to bat swing and throwing velocity of college softball players. 19 division I female college softball players were examined for anthropometric and performance variables. The anthropometric variables were percent body fat and lean body mass. The performance variables included grip strength, 1 repetition maximum (1RM) hang clean, parallel squat, bench press, and 1-arm dumbbell row, rotational power (4 lb medicine ball toss), vertical jump, peak power, throwing velocity and bat swing velocity. Prior to the day of testing, all female softball players practiced using the Myotest device to learn the protocol and the timing of the device. The Myotest device was attached to a neoprene belt that was wrapped around the softball player's hips. The softball player's place their hands on their hips, with feet apart and performed 5 countermovement jumps. The Myotest calculated vertical jump height, power, force, and velocity. Correlation analysis was used to explore the relationship among all variables. The results showed that there were significantly lower correlations between bat swing velocity and peak power, 1RM hang clean, body mass and percent body fat. Throwing velocity showed a low correlation to height and 1RM dumbbell row.

Parker et al. (2011) studied the effect of whole body vibration on lower body strength and power in the softball players. 9 NCAA Division-III female softball players were measured for 1- repetition maximum (1-RM) back squat, standing long jump and vertical counter-movement jump. The players were instructed to stand on the vibration platform for 30 sec prior to each set of squats. The vibration platform was set at a frequency of 35 Hz and at low amplitude. All the players were measured pre-test, at 4 week and 8 week periods. The results indicated that there were no differences between groups in percent change for back squat, standing long jump and vertical counter-movement jump. No differences were reported between groups and testing period in percent change for back squat, standing long jump and vertical counter-movement jump.

Shyamal Koley Et Al (2011) To verify the handgrip strength in Indian college softball players as dependent on decided on hand-anthropometric variables. Indian university softball gamers (121 adult males and 122 women) and untrained controls (98 men and 102 ladies) aged 18 – 25 years participated within the study. The following variables were recorded: body height and mass, BMI, right hand width and length, lengths of the right 2nd, third and 4th digits and right handgrip electricity. One-way ANOVA showed good sized differences.

Wong et.al. (2011) conducted a study to examine the repeated-sprint ability (RSA) in soccer players after 3 days of static stretching. Twenty soccer players (age: 16.8 ± 0.4 years) participated in 2 series of experiments with within-subject repeated-measure design (control series [CON]: 13-minute aerobic warm-up; and static-stretching series [SS]: 10-minute aerobic warm-up and 3-minute static stretching). Each series consisted of 5 days, and RSA (9 × 30 m separated by 25-second passive recovery) was tested on days 1 and 5. Static stretching was performed for 3 consecutive days from days 2-4, before and after intermittent aerobic endurance exercise on each day. The same warm-up protocol was used before and after all RSA tests and exercises within 1 series. No significant difference between CON and SS was observed ($p > 0.05$) in RSA for overall (all sprints), early phase (first to third sprints), middle phase (fourth to sixth sprints), and final phase (seventh to ninth sprints). Short-term static stretching had trivial effects (Cohen's $d < 0.35$) on overall and split RSA phases (early, middle, and final). The present study showed that performing static stretching for 3 consecutive days and before repeated-sprint test did not negatively affect RSA. However, it is premature to recommend that static stretching could be included in in-season daily warm-up routine because some movements such as jump and single sprint were more sensitive to static stretching.

Rodney j. Negrete et al (2012) Understanding the relationships among performance assessments and sport hobby is essential to the rehabilitation specialist. The Pearson product second correlation and a stepwise regression have been used to decide whether relationships existed between performance on the assessments and which upper extremity take a look at result pleasant defined the overall performance at the softball throw for distance. The results of this take a look at imply that several upper extremity checks display sizable relationships with one another and with the softball throw for distance. The changed pull up test changed into the first-rate predictor of overall performance at the softball throw for distance.

Szymanski et al. (2012) conducted a study to examine the effect of various warm-up devices on the bat velocity among the female softball players. 19 Division-I intercollegiate softball players of age 19.8±1.2 years were measured. The players had height 167.0±4.7cm, body weight 69.2±8.6kg, lean body mass 49.6±3.6kg and percent body fat 27.9±5.9. The players participated in a warm-up with 1 to 8 resistance devices on separate days. The players performed a standardized dynamic warm-up, 3 maximal dry swings similar to their normal game swing with the assigned warm-up device, 2 dry swings with a standard 83.8 cm, 652 gm softball bat followed by 3 maximal game swings while hitting a softball off a batting tee. The results revealed that there were no significant differences in bat velocity of softball players after using any of the 8 warm-up devices.

Nimphius et al. (2012) examined the relationship between percent change in muscle architecture and percent change in strength, speed and change of direction performance during the competitive season in female softball players.

10 female softball players of age 18.1±1.6 years from State Australian Institute of Sport softball team were measured for lower body strength, peak force, peak velocity and peak power. The players were also assessed for first base and the second base sprint performance, change of direction, aerobic capacity and muscle architecture characteristics of various muscles. The players were tested pre, mid and post training sessions. The results revealed that there were significant improvements in peak velocity and peak power. Significant increases also occurred pre-post in strength, change of direction on non-dominant side and second base sprint. There were strongest relationships between percent change in vastus lateralis muscle thickness and first base sprint, percent change in thickness and first base sprint, percent change in vastus lateralis fascicle length and second base sprint, and percent change in relative strength and change of direction on dominant side.

Koley and Kumaar (2012) estimated the grip strength of the dominant (right) hand and its associations with hand-anthropometric variables in 243 (121 males and 122 females) Indian inter-university softball players and compared them with 200 (98 males and 102 females) controls. The assessed anthropometric and hand-anthropometric variables comprised of height, weight, body mass index, shape index, digit index, 2D:4D ratio, palmar length, palmar width, palmar length/width ratio and right handgrip strength. The male and the female softball players were similar for age, the digit index and 2D:4D ratio and differed with respect to all the other study variables. The male softball players had significantly increased shape index, digit index, 2D:4D ratio, palmar width and palmar length/width ratio in comparison to their respective controls whereas female softball players had similar age, height and 2D:4D ratio to that in their parallel controls and differed for all other study variables. The coefficient of multiple correlations of height, weight, BMI and hand-anthropometric variables was observed to be lower in male softball players when compared to the female players though was higher than the controls. The right handgrip strength had significant positive correlations with height, weight, body mass index and all the six hand-anthropometric studied characteristics (except age and 2D:4D ratio in male players and age, shape index, digit index, 2D:4D ratio and palmar length / width ratio in female players).

Timothy j. Suchomel et al (2013) Strength asymmetry has been examined as a probable precursor of damage. Unfortunately, its effectiveness as an harm predictor is ambiguous. More these days, the feasible performance detriment associated with electricity asymmetry has been evaluated. This study separated athletes based totally on sex and no longer through game, which may additionally lead one to question whether the asymmetry is a result of intercourse or a by-product of unique game participation. Therefore, the motive of this examine turned into to examine bilateral electricity asymmetry among similar male and lady sports activities to decide if intercourse plays a role. A secondary reason of this observe changed into to assess the relationship of electricity asymmetry and mid high pull performance.

Thompson and Barkley (2013) evaluated changes in anthropometric and physiological parameters during a season among the division II collegiate softball players. 14 softball players were tested over the course of a complete academic year and competitive season encompassing 38 weeks. All the players were measured for body mass, body composition, aerobic capacity, anaerobic capacity, maximal upper and lower body power, speed, agility, and throwing velocity. The softball players were assessed at pre-season i.e. on 1st week, mid-season i.e. 15th week and post-season i.e. 38th week. The results showed that there were no significant changes in anthropometric characteristics among collegiate softball players evaluated at pre, mid and post season. Peak power was significantly improved among softball players at post season evaluation as compared to the pre and mid season evaluations. All other physiological parameters were not changed significantly among the softball players evaluated at pre, mid and post season.

Singh et al. (2013) compared the kin anthropometric and performance characteristics of elite and non-elite female softball players. 40 (20 elite and 20 non-elite) college level female softball players were assessed for height, weight, widths and diameters of body parts, girths, lengths and skinfold thickness. All female softball players were also measured for performance tests i.e. 50m dash, medicine ball throw, vertical jump, 10×4m shuttle run and reaction time. The results revealed that elite female softball players were found to have significantly greater height, biacromial, bi-illiocristal and bicondylar femur diameters and forearm circumference as compared to non-elite players. The non-elite female softball players were reported to have significantly greater thigh circumference, triceps and subscapular skinfold thicknesses as compared to the elite players. The non-elite players were also had significantly higher percent body fat as compared to the elite players. In performance tests, the elite players were found to have significantly better reaction time, speed and agility than the non-elite players. It is concluded that the elite female softball players had better kin anthropometric characteristics, body composition and performance characteristics than the non-elite players.

Kumar & Singh, (2014) conducted a study on comparative study of flexibility and leg strength between korfball and softball women players. The study included 60 subjects of which 30 were Korfball and 30 were Softball women players of Delhi and their age ranged from 17 to 27 years. Bent and Reach test was conducted to gauge the flexibility and Standing Broad Jump test was conducted to measure the Leg-Strength. The data was used to analyze by t-test. Significance was set at 0.05 levels. It can be concluded that there was no significant different between Korfball and Softball women players of Delhi. Korfball women Players having more Flexibility and Leg- Strength than the Softball women Players. Korfball women players performed better than the Softball women players of Delhi.

Hibberd, E.E. (2014) compared range-of-motion and humeral-retro torsion characteristics of collegiate baseball and softball position players and of baseball and softball players to sex-matched controls. Fifty-three collegiate baseball players, 35 collegiate softball players, 25 male controls (non-overhead athletes), and 19 female controls (non-over head athletes). Range of motion and humeral retro torsion were measured using a digital inclinometer and diagnostic ultrasound. Glen humeral internal-rotation deficit, external-rotation gain, total glen humeral range of motion, and humeral retro torsion. Baseball players had greater glen humeral internal-rotation deficit, total-range-of-motion, and humeral-retro torsion difference than softball players and male controls. There were no differences between glen humeral internal-rotation deficit, total-range of-motion, and humeral-retro torsion difference in softball players and female controls. Few differences were evident between softball players and female control participants, although range-of-motion and humeral-retro torsion adaptations were significantly different than baseball players. The throwing motions are similar between softball and baseball, but the athletes adapt to the demands of the sport differently; thus, stretching / strengthening programs designed for baseball may not be the most effective programs for softball athletes.

Pushkar Yadav (2014) The modern age of sports is the distinction, so in every sports activities perfection and purification of skill has got its sizeable importance. Biomechanics is an carried out shape of mechanics and consequently the technique used to analyze it should be derived from those of mechanics. Biomechanics goal to provide an explanation for the mechanics of life. Sports biomechanics have also generated different efforts at enhancing athletic performance in special video games and sports; it's miles helpful to decrease sports injuries through each identifying risky exercise and designing safer device and clothing. The purpose of this take a look at was to investigate the connection of decided on biomechanical variables with the hitting performance in softball.

Kumar and Singh (2014) conducted a study to compare the flexibility and leg strength between the korfball and softball female players. The study was carried out on 60 (30 korfball and 30 softball) female players of age ranged from 17 to 27 years from Delhi. Flexibility was assessed with help of bent and reach test and leg strength was measured by standing broad jump test. The results showed that there was no significant different between korfball and softball female players of Delhi on variables flexibility and leg strength.

Singh and Gaurav (2014) compared the hand grip and shoulder girdle strength among cricket, baseball and softball players. 36 male college level players (12 cricket, 12 baseball, 12 softball), were assessed for height, weight, hand grip strength and shoulder girdle strength. Grip strength was measured with the help of hand dynamometer and shoulder girdle strength was measured by medicine ball throw. The results showed that there were significant differences between cricket, baseball and softball players with regard to hand grip strength and shoulder girdle strength. It was further revealed that baseball players had better right and left hand grip strength and shoulder girdle strength than the cricket and softball players.

Singh (2014) compared the physical fitness and psychological variables of softball and cricket players. 20 softball and 20 cricket players were tested for agility, arm strength and depth perception. Arm strength was measured with the help of softball throw test. Shuttle run was used to measure agility. Depth perception was measured with automated vision screener. The independent samples t-test revealed that there were no significant differences in agility, arm strength and depth perception between the softball and cricket players.

Sharma et al. (2014) compared the physiological and physical characteristics between inter-university and inter-college male softball players (n=30) from Punjab. The players were divided into two groups, one comprising those who participated in inter-university championship at least one time in last two years (n=15) and the other comprising players who participated in inter-college tournament but never got selected for any inter-university championship (n=15).The blood pressure and pulse rate were similar in both the groups though there was a significant increase in explosive power of inter-university in comparison to the inter-college group.

Nathan j. Hoffmeier (2014) This study is an attempt to provide assist for the medical use of a unmarried session Graston Technique intervention to enhance flexibility and decrease musculotendinous stiffness (MTS). Thirty subjects have been recruited to participate inside the have a look at from the Oklahoma State University Athletic Training Education Program and related Master's programs. However, subjects gravitated closer to baseline 48-hours put up-intervention. Results of this have a look at recommend that a unmarried-session GT intervention would present no huge brief-term or long-term outcomes on MTS of flexibleness.

Abd Rahim Bin Mohd Sharif Et Al (2014) The number one objective of the existing paper offers with to examine the connection of hockey coaches and crew cohesion with the performance of field hockey gamers of Pakistan and their effect on gamers' performance. However the secondary goal was to degree the sphere performance of hockey gamers on practical measures. The studies method is based on both descriptive and inferential statistical methods. The descriptive statistics was amassed inside the shape of discipline overall performance checks (technical competencies and fitness capabilities) whilst the inferential information perceived via gamers became collected using survey questionnaire. A wide variety of 296 national discipline hockey gamers of Pakistan had been decided on from 14 countrywide departments of area hockey as samples.

Bhagat et al. (2015) compared the anthropometric, physical fitness and psychological characteristics of softball and cricket players. 30 softball and 30 cricket male school level players were measured for BMI, upper arm, forearm, hand, leg and foot lengths, speed, flexibility, agility, strength and depth perception. The results revealed that the cricket players had significantly greater hand length, leg length, and strength in comparison to softball players. On the other hand, softball players were significantly better in speed and agility than the cricket players. No significant differences were observed in all other variable among the two groups.

Singh & Singh, (2015) assessed the study of a certain relationship of anthropometric and body structure and composition characteristics with the batting skills among softball players. For the above case study around 150 male players aged between 18- 30 years from the different states and universities of India were selected to participate in the study. The various aspects were measured as follows; body weight was measured by usage of digital weighing machines. Height was measured with the help of flexible and tape. The width and length of the body were measured with the help of a sliding caliper. The thickness of the skin fold was assessed using the measuring device known as Harpenden skin fold caliper. The body fat percentage was estimated from the sum total of skin folds calculated using the standard equations. The Karl pearsons product moment analysis was performed to know the facts of anthropometric and body composition parameters with the batting skill in the softball players. The following results of correlation analysis revealed a specific association of weights (p==0.009), total arm length (p=0.032) and the lower arm lengths (p==0.043) with the batting ability. The batting skills were also diligently observed to be specifically associated with upper arm circumference (p==0.004), foreman circumference (p=0.004) and chest circumference (p==0.011). The biacromiobn (p==0.003) and the bicondylar humerus (p4) and the overall chest circumference (p==0.011), and also the thickness of the skin fold were noted. The research shows that the anthropometric variables had an important role in the performance of softball players.

Sintara and Sonchan (2015) examined the physical fitness of the collegiate male and female softball players from Thailand. 15 male and 15 female softball players aged between 18-22 years were assessed for health-related fitness and skill-related fitness. The health-related fitness includes variables such as grip strength, leg strength, percent body fat, flexibility and vital capacity whereas the skill-related fitness included power, agility and base running. The physical fitness data of the players was compared with the data of collegiate softball players of US. The averages of the base running speed were lower than the collegiate softball players of US. The health-related fitness was lower than the standard data of Sports Authority of Thailand.

Wentzel and Travill (2015) conducted a study to determine the relationship of morphological and fitness characteristics with the skill and performance measures in men's fast-pitch softball players. 15 provincial and 15 club male players were measured for anthropometric characteristics, fitness tests and skills of softball. The results showed a number of strong significant relationships between the different variables tested. Height and percentage body fat were found to significantly associated with the two key performance areas, batting and base running.

Fajar Awang Irawan (2015) The good throws Want acceleration and acceleration starts off evolved from most lateral rotation with the ball come fast to the target. The purpose of this take a look at turned into to investigate the assessment among sidearm and overhand throws. Base on the movement evaluation view in Dart fish, sidearm throw is less difficult and faster, which meant aspect arm throw was more easy to apply than overhand throw by means of the players. Sidearm throw greater powerful than overhand throw, due to the fact the movement is shorter and less difficult.

Upma Bhagat Et Al (2015) The reason of this study is to compare decided on Anthropometric, Physical fitness and psychological variables between Softball and Cricket male players at country degree. The take a look at became delimited to following variables: Anthropometric variables: frame mass index, higher arm duration, fore arm duration, hand length, leg duration and foot period. Physical fitness variables: Speed, agility, flexibility and energy. Depth notion. To compare decided on Anthropometric, Physical fitness and psychological variables among Softball and Cricket male gamers at nation level imply, fashionable deviation and t-take a look at have been employed with the help of statistical package deal of SPSS. To check the hypothesis the importance level becomes set at 0.05 percent.

J Matt Mccrary Et Al (2015) This systematic review became carried out to discover the effect of upper frame heat-up on performance and harm prevention results. Inclusion standards have been English language randomised controlled trials from peer-reviewed journals wherein investigation of higher frame heat-up on overall performance and damage prevention outcomes was a primary purpose. Conclusions Strong research-based proof became observed for the following: high-load dynamic power and energy performance; warm-up swings with a standard weight baseball bat are simplest for reinforcing bat velocity; short-duration static stretching warm-up has no effect on strength consequences; and passive heating/cooling is a largely ineffective warm-up mode. A clean understanding gap in higher body warm-up literature is the dearth of research of harm prevention final results.

Rao and Kumar (2015) conducted a study to examine the relationship of physiological parameters with the playing ability among the softball players. 50 under 15 softball players were assessed for physiological parameters viz. resting pulse rate, blood pressure, vital capacity, percent body fat, anaerobic power and respiratory rate. The correlation analysis revealed that the only respiratory rate was found to have significant relationship with the playing ability among the softball players. The multiple regression analysis showed that the only respiratory rate significantly predicted the playing ability among softball players and accounted for 6% of total variance. All other physiological parameters were not significantly contributed to the playing ability among the softball players.

Kakran, (2016) evaluated a study on comparative study between cricketers and male softball players on coordinative abilities. The study comprised of forty (N=40) male subject aged between 18-25 years, who had participated in interuniversity competitions from Jiwaji University. The technique used to attain the objectives of the study was Purposive sampling technique. The subjects were further divided into two groups, N1=20 cricket players, N2=20 softball players. T-Test which is independent was applied to find out the significant difference of coordinative abilities between the two groups N1 and N2. Level of level of significance was set at 0.05 to test the hypotheses. It was concluded from the results obtained that there is a significant difference between cricket players and softball players on various sub variables that include reaction ability, orientation ability and differentiation ability. However, there was not much of a difference observed with regards to one of the sub-variable rhythmic ability.

Sathuluri et al., (2016) conducted current investigation to find out the relationship between prime parameters of motor fitness and anthropometric parameters with playing ability of softball players. It included the subject of 50 softball players who participated in All India Softball Ranking tournament in under ''15'' category. The age of the group ranged between 14 and 15 years. The study was all about the abdominal strength and trunk length and therefore multi regression study is used to conclude the study. To achieve the study to be more effective, they collected all the specific data, orientation procedures have been presented. Present study to find out the relationship between selected motor fitness variables with playing ability, the multiple regression analysis, the motor fitness variables abdominal strength, were found to be significantly related with playing ability with obtained 't' values of 3.494, 2.069, 2.522 and 3.626 as these 't' values were greater than the required table value to be significant at 0.05 level. It was concluded that motor fitness variables abdominal strength were significantly related with playing ability of softball players. It was concluded that Anthropometric variables Trunk Length were significantly related with playing ability of softball players.

Hitoshi Shitara Et Al (2016) We prospectively evaluated the results of a prevention program on the occurrence of shoulder and elbow injuries in high college baseball pitchers. Ninety- pitchers participated on this study and have been taught to carry out stretching and energy sporting events aimed at enhancing shoulder outside rotation energy in the preseason. The pitchers freely chose to participate in considered one of 4 agencies [SM-group:

performed both exercises, S-group: performed stretching exercise only, M-group: performed strength training only, and N-group: performed neither intervention]. Injury became defined as lack of ability to play for ≥8 days because of shoulder/elbow symptoms. Daily posterior shoulder stretching may lessen the prevalence of the injuries in excessive college baseball pitchers.

Sang-Won Lee Et Al (2016) The purpose of this study became to evaluate the effect of thoracic dynamic stretching with elastic band on the ROM and ball velocity for baseball participant. For this cause, 11 baseball gamers of the I town high faculty and college done thoracic dynamic stretching with elastic band earlier than throwing a ball. After stretching, we measured the ROM and ball velocity and as compared. The results were as follows. First, ROM after stretching showed a great increase straight away in comparison with earlier than stretching. Second, the speed of the ball was significantly expanded after stretching compared with before. These consequences indicated that thoracic dynamic stretching with elastic band for baseball gamers would possibly have a superb effect on ball speed as well as ROM.

Chhina, (2017) to evaluate the relationship among the body compositions to throwing performance of softball players. For the study 45 softball players were taken who represent their college in inter collegiate competition. Age of the players is between 17 to 19 years. Anthropometric variables like height, weight and skin fold thicknesses were assessed. AAHPERD softball skill test battery was used to assess throwing skills of the softball players. To test the statistical data the Karl Pearson's co-efficient of correlation was computed to assess the relationship between body compositions to the throwing performance. It was concluded that height, weight, BMI and skin fold thickness shows there a negative relationship with throwing performance among softball players

Ryan r. Fairall et al (2017) To study the acute effects of self-myofascial release (SMR), static stretching (SS), and the mixture of self-myofascial launch and static stretching (SMR þ SS) on glen humeral inner rotation range of motion (GH IR ROM) in overhead athletes with glen humeral inner rotation deficit (GIRD). Participants: Twelve asymptomatic grownup male novice softball function players displaying GIRD. Results: All 3 methods considerably progressed GH IR ROM. Post hoc checking out found out that SS by myself and SMR þ SS stepped forward GH IR ROM considerably more than SMR alone. However, there have been no big variations in GH IR ROM between SS on my own and SMR þ SS.

Kanwar Mandeep Singh Et Al (2017) The goal of the existing study was to evaluate the relationship between the anthropometric variables and throwing skill in among male softball gamers. Total one hundred fifty male university and national level softball gamers of various universities and states of India had been decided on thorough purposive sampling technique. The age of subjects ranged between 18 to 30 years. Body weight of the subjects changed into measured with a portable weighing machine. Height measurements were taken the usage of the usual anthropometric rod. Body mass index become then calculated using the components weight (kg)/height2 (m).

Lane b. Bailey et al (2017) Baseball players displaying deficits in shoulder variety of motion (ROM) are at expanded danger of arm harm. Currently, there is a lack of consensus concerning the excellent to be had remedy alternatives to restore shoulder ROM. Instrumented manual remedy with self-stretching will result in clinically significant deficit reductions when compared with self-stretching on my own. Deficits in inner rotation, horizontal adduction, and general arc of movement have been as compared between corporations without delay before and after a single remedy consultation. Instrumented manual therapy with self-stretching extensively reduces ROM chance factors in baseball players with motion deficits whilst as compared with stretching alone. Clinical Relevance: The brought blessings of manual remedy might also help to reduce ROM deficits in medical scenarios wherein stretching on my own is useless.

Koya Mine Et Al (2017) This systematic review aimed to accumulate and synthesize effectiveness information from English- and Japanese-language randomized managed trials (RCTs) investigating stretching interventions for PST and GIRD. Cross-body stretch may be powerful to improve PST and GIRD in asymptomatic young subjects at once or in the quick time period. Further take a look at with methodological rigor is necessary to research the long-term effectiveness of stretching interventions on PST and GIRD in symptomatic patients.

Daksh Sharma (2017) To satisfy the motive of these have a look at total sixty girls among the age institution of 18-two decades had been decided on as subjects for the study. Out of those 30 ladies was Softball gamers, who participated at College level competition and different thirty women were Non-gamers, who in no way took lively component in any occasion or game at any age degree. The purpose of the study changed into to examine the anthropometric variables of Softball gamers and Non-Softball players. Means, fashionable deviation and independent student's test have been used as statistical equipment. It was determined that Softball gamers have been barely heavier and taller than the Non-Softball players, but the variations were Non-huge.

Kanwar Mandeep Singh (2017) The aim of this study became to check relationship of physiological parameters with playing potential amongst softball players. Total a hundred and fifty male softball gamers from the numerous states and universities of India had been selected to take part within the take a look at. Vital capability of the subjects became measured with the assist of the automatic spirometer. Cooper's 12 mins run/walk check turned into used to assess the aerobic fitness (VO2max) of the topics. Speed turned into measured with the 50m dash. Explosive power of the subjects changed into determined through the usage of the standing vertical leap. Sit and attain test changed into used to measure the ability of the subjects.

Kanwar Mandeep Singh (2017) The reason of this examine was to envision dating of physiological parameters with fielding talent amongst softball gamers. Total 150 male softball gamers from the diverse states and universities of India had been decided on to take part within the take a look at. Vital potential of the subjects become measured with the help of the automatic spirometer. Softball gambling ability of the players become assessed by means of AAHPERD softball skill take a look at battery. The results found out that the physiological parameters i.e. Cardio fitness (VO2max), explosive energy, speed, flexibility and grip energy of each proper and left hand established a giant correlation with the fielding ability a few of the softball gamers.

Kanwar Mandeep Singh Et Al (2017) The intention of the take a look at became to ascertain the association of physiological parameters with the throwing performance many of the male softball gamers. Physiological parameters particularly important capacity, cardio fitness (VO2max), velocity, explosive power, flexibility and grip energy have been assessed the use of standardized checks. Throwing skill of the gamers was assessed with the AAHPERD softball talent take a look at battery. Karl Pearson's product moment co-green of correlation became computed to assess the relationship between physiological parameters and throwing ability test of softball some of the softball players. Significance degrees have been set at.

Singh et al, (2017) studied the relationship of physiological variables to the throwing skills of softball players. The sample of 150 male University players from different Universities of India has been taken and their mean age was 20.89±1.54 years. Aerobic fitness, vital capacity, speed, explosive strength, flexibility these are the physiological variables used to standardized tests. To test the statistical data the Karl Pearson's co-efficient of correlation was used for the present work. Present study reviled that aerobic fitness, speed, explosive strength, flexibility and grip strength had a positive relationship where as vital capacity has negative correlation among the throwing performance of softball players.

Aswathy, (2018) compared the hand grip strength with hand anthropometric variables of inter-university level Softball, Cricket and hockey players. The study included 45 inter university sports person aged between 18-25yrs. The subjects included were 15 softball, 15 cricket and 15 hockey female players. A total of 13 variables were used which included 11anthropometric variables and 2 strength variables (Grip strength left, Grip strength right). The data were collected using standard anthropometric equipment (Gulictape, sliding caliper, skin fold caliper, and grip dynamometer) which were supplied by reputed Indian and foreign companies and used in the sports science laboratory of Lakshmi bai National college of Physical Education. The outcomes of the study revealed that the grip strength of softball players showed the strongest relationship with the arm measurements. Cricket and hockey players correlated second and third, respectively. It was concluded that the grip strength of University level women softball players had considerable relationship with most of their arm measurements in comparison to cricket and hockey players.

Singh, (2018) studied the relationship of anthropometric variables and body composition with the performance level among softball players and to locate the anthropometric indicators of performance in softball. A group of 150 male softball players from different states and colleges of India were chosen and examination for the study and the mean age of the group is 21.35±2.57 years. Height and length estimations of the players were estimated with the anthropometric pole. Versatile gauging machine was utilized to quantify the weight of the players. Circumferences and distances across of the body portions of the players were estimated with steel tape and sliding caliper individually. Harpenden skin fold caliper was utilized to survey the skin fold thicknesses. The results of the study showed that the height ($p=0.010$), weight ($p=0.001$), absolute arm, upper arm and lower arm lengths ($p<0.05$); Upper arm, lower arm, chest and calf circuits ($p<0.05$); biacromial, bicondylarhumerus and bicondylar femur measurements ($p<0.05$) and fit weight ($p=0.000$) were altogether connected with execution among softball players. Relapse examinations demonstrated that for softball execution, 13% fluctuation was represented by slender weight, 3% was clarified by upper arm perimeter, and an extra 3% of the change was clarified by upper a careful distance. Result concluded that lean body mass was the principal indicator of the softball performance.

Ibrahim Hamed Ibrahim Hassan (2018) The aim of current study Changed into to analyze the relationship between strength and linear or non-linear dash overall performance variables in U19 male area hockey players in the course of the midseason time. In addition, no big relationships located with different dash distances or change of path measures while correlated to different variables of VJ and energy results. They have a look at concluded that, the information of relationships among strength and cyclic or acyclic sprint performances at diverse distances could allow coaches to shape the sphere hockey schooling drills.

Kelly p. Massey (2018) This observes examined the time difference between three one of a kind base strolling begins strategies. Thirteen Division II collegiate softball gamers accomplished maximal sprints off a softball bag at two extraordinary distances. Results recommend that the rocking start technique be carried out because the beginning functions off the bag in fast pitch softball. Implementing the rocking approach might also decorate performance and success in stealing and base going for walks.

Marquardt et al., (2018) conducted and examination to inspect the impact of various beginning stances on acceleration and speed stages in university softball players. Seventeen solid NCAA Division I women's softball players (age = 19.9 ± 1.3 yrs) elected to take an interest. Three most extreme 45 ft sprints, with 1-minute rest, were performed (with parts at 15, 30, and 45 ft) for every one of three starting stances. A 1x3 repeated measures ANOVA for complete-time exhibited that front foot on the base was fundamentally quicker than a back foot on the base and the traversing step for each of the three parts, front foot on the base was likewise essentially quicker than the back foot on the base and traverse step. The decrease in time for the front foot on the base was presumptively the consequence of utilizing the base to push against, almost like a runner's square, to deliver more prominent even power to quicken quicker and arrive at a more noteworthy top speed. Hence concluded, mentors should show their softball competitors to remain with their front foot on the base when base running.

Kobak et al., (2018) carried out a study to find the effects of medicine ball training on bat swing velocity in prepubescent softball players. The study included 27 subjects who were all female and prepubescent softball players with an average age ranging from 10.2 ± 1.2. The subjects were divided randomly into two group's one group for Medicine Ball training (13subjects) and the other group for Control (14subjects). The test included 10 dry bat swings and followed 10 hits off of a pitching machine with bat swing velocity measured for each swing using Swing Speed Radar Measurement Device. The Medicine Ball group completed sport-specific medicine ball throws that included side medicine ball throws, Russian twists, woodchoppers, and standing band rotations aimed at developing rotational velocity. From the study, it was concluded that there were no significant interaction effects between both medicine Ball and Control Group conditions from pre- to post-testing for average bat swing velocity. As a result, medicine ball training offers no additional benefits in bat swing velocity.

Nicholas j. Kelling et al (2018) Batting in baseball or softball represents a physically and perceptually challenging assignment. Changes in flight of a high-velocity pitched ball require brief and accurate predictions of future area. To achieve success, an individual must be capable of swiftly acquire and process visual information, suggesting an emphasis on spatial working memory. The contemporary test assessed if people of variant knowledge tiers (novices and varsity softball players) differed in ability to determine destiny places of a pitched ball primarily based on special pitch types and durations. Data propose an impressive base capability for visible movement prediction which includes a time suitable ability to be expecting motion timing.

Additionally, while not primary to this functionality, data suggest relevance for spatial working reminiscence in predicting speed.

Amy Marquardt Et Al (2018) Speed is a vital aspect in softball, and may be the difference among prevailing and dropping. Base stealing is a way used to provide runs. There has been debate over which starting role is the most effective to maximize acceleration and pace to attain the next base the quickest. The reason of this observes become to study the effect of different starting stances on acceleration and pace levels in collegiate softball players. The decrease in time for the front foot on the base changed into probably the result of the use of the base to push against, like a sprinter's block, to produce greater horizontal force to boost up faster and reach a extra top pace. Coaches need to educate their softball athletes to stand with their front foot on the bottom while base strolling.

Saraya et al., (2018) team determined the factors and Physical Conditions which are dominant in determining the batting skills in softball. Simple random sampling technique was used to sample out 100 subjects for the study. The data applied are carried across a sequence of tests of the independent and dependent variables. The test data were analyzed using SPSS, along with confirmatory factor KMO and Bartlett's Test. It was concluded that factor of anthropometry dominant physical condition such as arm power, arm length and reaction time determined the skills of a batter in game of softball.

Venkatappa & Viswanath, (2018) conducted a study to find out the difference level of physical fitness among cricketers and softball players. The sample of 120 players in which 60 from cricket and 60 from softball has been taken from Dr. B R Ambedkar UG and PG College, Baghlingampally and their age range from 18 to 24 years. Physical fitness components like leg power, arm strength, speed, flexibility and endurance have taken into consideration for the present study the level of difference among cricketers and softball players. In this study the data was collected to use the various test items which taken from different physical fitness batteries. The data was composed , leg power was measured by standing broad jump, arm strength was measured by using 6Lbs medicine ball put test , 50 yard dash test was measured for speed, flexibility was measured by Kraus-Weber floor touch test and 12 min run and walk trial was use to assess endurance. To compute the raw data mean, standard deviation and Z ration were employed for statistical analysis. Results shows that both cricketers and softball players have shown similar level of performance in trunk flexibility and endurance and cricketers have better performance in three physical fitness components like Leg power, Arm strength and speed.

Akbar & Khan , (2019) studied prominent Anthropometric and Physical fitness factors among national level men softball players in Kerala. A total of 40 men softball players with age ranging from 17-25yrs were selected for the study. Twelve Anthropometric variables which include Height, Weight, Sitting height, Leg length, Arm length, Thigh girth, Calf girth, Ponderal Index, Crural Index, Body mass index, Biceps circumference and Mid forearm circumference were considered for the study. The Physical fitness tests included Speed (40mts flying start), Endurance (1200 mts), Agility (4x10 shuttle run), Flexibility (sit and reach), abdominal strength (sit up in minutes), static balance, leg explosive strength (standing broad jump), hand explosive strength (cricket ball throw). It was conclude that anthropometric variables like weight, girth, forearm circumference, leg length and in physical fitness variables leg explosive strength, abdominal strength and cardio vascular endurance were more prominent among men softball players. Based on the study, the recommendations include anthropometric and physical fitness variables may be conducted using university level men softball players, similar studies may be conducted using female Hockey players of other states and similar study may be conducted using female softball players.

Rafael e. Reigal (2019) The first objective of these studies changed into to analyze the relationships among the sports mental profile, competitive tension, mood and self-efficacy in beach handball games. The second objective became to decide the predictive ability of the mental profile on aggressive tension, moods and self-efficacy, which became accessed via linear regression evaluation. The Psychological Sports Execution Inventory (SPPI), the Competition Anxiety State Inventory 2 (CSAI-2), the Mood Profile (POMS) questionnaire and the General Self-Efficacy Scale (GSES) had been used to attain the information. Correlation and linear regression analyses monitor statistically sizable institutions among the constructs studied, each for the full pattern and by means of gender.

Mandeep Singh Et Al (2019) The reason of this examine become to study body composition additives amongst softball players and to compare body composition components among male and lady softball gamers. Total one hundred fifty male and one hundred fifty girl softball gamers from the various states and universities of India have been selected to take part in the study. Height of the gamers turned into measured with the anthropometric rod. Portable weighing system became used to measure the load of the gamers. Harpenden skin fold calliper becomes used to evaluate the skin fold thicknesses.

Hallvard Nygaard Falch Et Al (2019) The capacity to carry out a rapid exchange of course (COD) is a essential skill in several court- and field based sports. The goal of this evaluate is to analyze the impact of various bodily education bureaucracy on COD overall performance. A systematic evaluation of the literature turned into undertaken the use of the following databases: Pub Med, SPORT Discus and Google Scholar. Studies have been eligible if they met the following standards: (1) a COD check measuring overall performance before and after the schooling intervention, with unique description of the take a look at in terms of length and number of changes in a direction with specified angles, (2) involve schooling intervention like plyometric, power, sprint, particular COD education, or a combination of those education paperwork targeting the decrease extremities, (three) the take a look at needed to country training historical past in phrases of which game they participated in and their aggressive level and an in depth methodological description.

Jung-Hyun Choi (2019) The purpose of this observe turned into to evaluate the effects of proprioceptive neuromuscular facilitation (PNF) with elastic band workout on strength and dynamic balance in ankle instability. A general of nineteen topics with ankle instability. Subjects have been randomly divided into ST(Stretching group, n=10) and STPNF(Stretching with Proprioceptive Neuromuscular Facilitation institution, n=10). Two agencies measured power dynamic balance. The differences in ankle muscle strength and balance capacity between pre- and post-intervention have been analyzed the use of paired t-test at and impartial t-test.

Rozella ab. Razak etal (2019) One of the most important competencies in softball is throwing. Various methods of education have been tested to determine an alternative method that could improve throwing ball speed. Combined training is one method that these days won interest among sports activities practitioners. However, maximum training focuses handiest at the main muscle tissues and neglects assisting muscle tissues together with handgrip and trunk rotation muscle tissues which can be also crucial in the execution of throwing. Therefore, this take a look at seeks to determine the impact of handgrip, trunk rotation, and mixture of handgrip and trunk rotation training at the throwing ball velocity of collegiate female softball gamers. A total of 72 collegiate lady softball gamers have been assigned into four electricity education corporations specifically handgrip (HG), trunk rotation.

Nasreen Ashra Et Al (2020) Background: Badminton is a sport that calls for quite a few overhead shoulder motions. Abduction, inner and external rotation of the shoulder joint in the course of photographs, lobbing, stroking and smashing were the cause of the injury. These alterations were related empirically to bony and gentle tissue diversifications that result from the large rotational and distractive forces acting on the GH joint all through the throwing movement. Study subjects had been evaluated before mission their assigned stretching classes for their shoulder internal rotation ROM, using a mechanical inclinometer. Sleeper stretch turned into repeated three instances and held for

30 seconds with 30 seconds of relaxation duration. PNF stretch became carried out 3 instances with 20 second of rest length in among every stretch.

Shrikant S Mankar (2020) The purpose of the observe was to discover the role of yogic sports on Softball players. This have a look at may additionally offer a standard guideline for sportsperson. By adopting yogic way of existence overall performance may be developed. The yogic exercises improve their intellectual and bodily health, reduces stress and anxiety, calms the mind, improves balance, presents flexibility within the arms, hands, wrists and fingers. The yogic sports help in retaining the backbone erect, beneficial in constipation and improve digestive power. It is beneficial in arthritis, it improves agility and strength. In the stop of the look at we will say that yogic physical activities results more on Agility and Strength of Female Softball Players.

2.3 Summary of Related Literature-

The review of above literature has given a lot of insight into the present study. The research studies undertaken across the globe about the performance predictors or factors influencing players' performance indicated that there are various types of variables that could be considered. They ranged from physical fitness and psychological variables to physiological of the players. Though there is considerable diversity in the kind of inferences drawn, the first three variables groups have been subject of investigation for most of the researchers.

Further the variables considered in each of the groups mentioned above varied substantially. Keeping these factors in mind, the important variables were finalized for the study. The review of literature also helped in finalizing the research objectives, hypothesis and limitations of the study. In the light of review of related literature the methodology followed in the present study was finalized and the same is presented in the next chapter III.

CHAPTER-III
METHODOLOGY

3.1 Introduction-

The present study was designed to selection of topics, selection of variables, reliability of the statistics, reliability of questionnaire, orientation of subjects, administration of questionnaire, collection of statistics and statistical techniques employed for studying the facts have been defined. The Study became designed to deal with the comparative analysis of effectiveness of stretching on performance of the softball players variables and socioeconomic elements many of the university Softball gamers at one of a kind in India.

The research scholar reviewed the available scientific literature, journals, periodicals, magazines and research papers maintaining in thoughts the role of decided on mental variables and socio competitively priced factors, their importance, availability of questionnaires and feasibility components in their measurements, number of variations have been selected.

The studies are a systematic method of collecting, recording and analysing of records. The present research work turned into of descriptive kind and offers with courting among variable, hypothesis trying out and development of generalizations, principles and theories having familiar validity and is worried with useful courting. Descriptive studies are not restricted to simplest one technique of facts series, these might also appoint any or all the strategies like, observation, questionnaire, interviewing, scaling techniques and so on. The essential method of this research was the survey method which became complemented by using in-depth interviews and the evaluation of files.

The statistics amassed became analyzed the usage of general statistical tools to attract inferences and the outcomes received after evaluation had been provided in graphical presentation. For analysing the facts percentage become used as statistical technique. Hence taking in consideration all these data the present researcher conducted the study on comparative analysis of effectiveness of stretching on performance of the softball players.

Qualitative and Quantitative method has been used to do research for this look at. Qualitative researches are concerned with context due to assumption that human behaviour is drastically influenced by means of the putting in which it occurs. Therefore, qualitative researchers choose to visit the area of the problem in their studies. As the real findings of the examiner may be beyond anticipation of the researcher, qualitative technique for this study will try and find out the solution of the research questions. This study specializes in the current troubles of sports activities accidents in softball and is expected to be inductive and exploratory in nature. In qualitative study there is flexibility in which researcher enjoys freedom in using theories and speculation to exchange as event unfolds. As the survey became related to Softball participant, pupil has taken large efforts to accumulate most reviews, non-public observations and perspectives accumulated by using the scholar at some point of survey helped research

Softball is a popular game played throughout world. Very few studies were undertaken in the field of softball to study different factors, which continue towards successful performance. There is a paucity literature in softball and several areas require research and investigation.

3.2 Design of Study -

Research refers to a systematic process which is completed in step by step. Before going to the actual research work, researcher plans his entire programmed of research. This is called research proposal Research proposal is a concept base structure. Tabulation helps for collection of data, interpretation of data, measurement of data research proposal is most important. The proposal is comparable to the blue print that an architect prepares before the bides were let and building commences.

3.3 Method -

Good research is the result of using right methods and tools. The success of any research work merely depends upon the instruments selected for the study. A research technique as such, can be into the showing method-

(1) Historical Method– The past study is termed as the basis and the future study is done on the basis.

(2) Descriptive or Survey Method– It is concerned with the present and on its basis the standard of the study is ascertained.

(3) Exploratory or Experimental Method– The aim of such type of study is to show the relationship between two or more matters and to explain the relationship precisely.

In the present study the survey method has been used. Survey method is the most commonly used approach for data collection. In this method, random sample of participants complete a survey test. Survey method is fast, cheap, easy and most flexible than other methods. Researchers can collect large amount of data this technique. The description of this method is described below:-

3.3.1 Survey Method -

Researcher has used survey method to get the information of the present circumstances. In this method explained comparison measurement, classification and generalization etc. are used. Survey method is mostly used in the educational researcher. A survey may be inspection or observation of an object.

Essential Feature of survey method -

1. The data can be collected about many persons simultaneously in this method.
2. By this method the survey of individual or objects may be done independently.

3. This method is concerned with some geographical fact or the study of a place.
4. This method helps in the understanding of internal and external aspects of an individual.
5. This method may be qualitative or quantitative or both.
6. The complete plan of a research or the survey of the steps of most area of a research can through survey method.
7. This method includes the formation of the activities relevant to a group and the different aspects implicit with them.
8. The define problem and definite aims are included in this method. The explains and analysis of the facts and deriving the results necessitates logical and expert faculty in a researcher.

3.4 Population of Study -

In the field of physical education, the population under study is quite large, which is practically impossible to contact and study individually. Hence, sample of the total population to be investigated is selected randomly with assumption that they truly represent the total sample. The population of study is present study were delimited to Softball Players who were the participants of inter-collegiate level tournament and inter-zone softball championship who represent respective colleges affiliated to various Universities in Rajasthan state and also belong to rural and urban area from Rajasthan state. From this vast population 170 softball players of inter-collegiate level of Rajasthan state were randomly selected.

3.5 Selection of Subjects-

For the study researcher has selected one hundred seventy (N=170) male & female collegiate Softball players who were participated in the Inter Collegiate and inter-zone softball championship who represent respective colleges affiliated to various Universities in Rajasthan state and subject aged between 20 to 30 were selected as subjects for the study.

The full table of sample is shown given below-

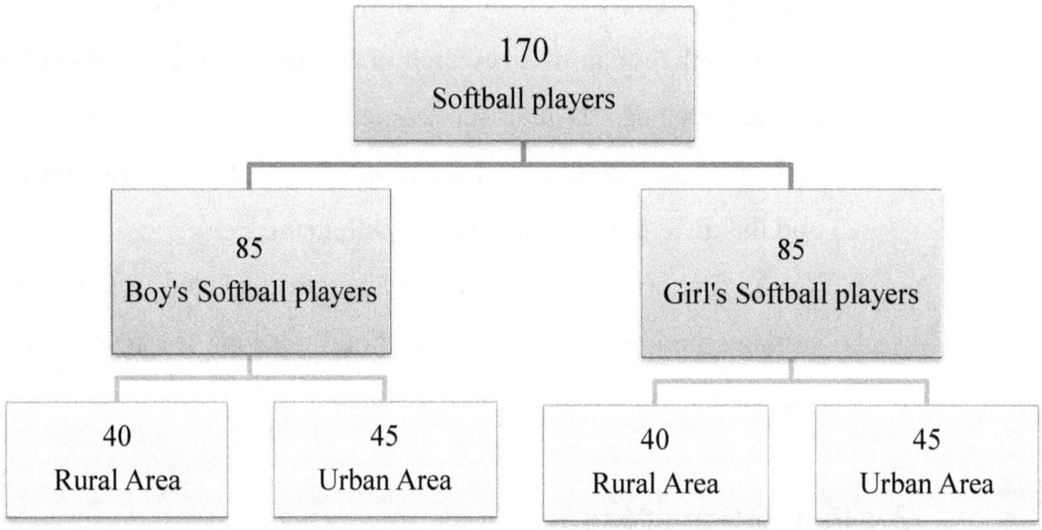

3.6 Selection of Variables-

The researcher reviewed number of research articles, coaching manuals, journals and books through the scientific literature pertaining to the analysis of performance measurements, physical fitness and stretching variables from different sources and also consulted the experts in these areas. Along with the said literature and expert opinion, keeping in mind the objectives of the study and availability of standard instruments for measurements and recording of data was also taken into consideration while selecting performance measurements, physical fitness, and stretching variables. Based on these observations, the investigator selected the following variables were selected for softball players assuming these variables might be significant for optimal performance in Softball.

3.6.1 Dependent and Independent Variables-

- **Dependent Variable-**

Performance-

Total performance of the selected subjects will be rated by experts subjectively on a 10 point rating scale (Coach Rated Performance Evaluation). This will be the dependent variable for the study.

- **Independent variables-**

The independent variables are Stretching and Physical Fitness variables. Specialization of measurement and the equipment to do so was found necessary.

The researcher selected stretching exercises as the independent variable for this study. In softball, lower body strength is required for kicking, jumping, tackling, twisting and turning and also forms the foundation for explosive speed. Upper body strength is required for shielding the ball, holding off opponents, throw-ins and also contributes to overall power and explosiveness. Strong legs improve the ability to play softball. Strong legs carry the player as the player run up and down the field. The stronger the muscles in the lower body, the stronger the ability to run, kick, shoot and tackle.

Stretching exercises are therapeutic manoeuvre that are done to the skeletal muscles. Wherein the origin and insertion are moved away from each other with respect to the normal joint range possible stretching exercise are beneficial in improving the flexibility of soft tissue around a joint. An improvement in the flexibility will have a direct effect on the performance. Physical requirements of softball players have to make repeated dynamic movements, such as sprints, jumps and turns. These movements all require dynamic muscular contractions; contractions that have little relevance to those involved in held stretches. Most top clubs now employ dynamic warm-ups, which place a much greater emphasis on active and football relevant dynamic mobility and less cause of injury.

Training schedule-

The purpose of the study was to find out the effect of stretching exercises on the performance of softball skills of collegiate players. To achieve this purpose of the study 70 softball players from various colleges in Rajasthan state, India were selected in the age group of 20 to 30 years.

Table No. 3.1: Selected Stretching Exercise, Tests and Criterion Measures

Exercises Groups	Weeks	Days (alternative)	Session	Time	Method	Rest Days
Stretching Group	8	3	Evening	4.30 PM to 5.30 P.M	Warm-up Exercises Warm-down	4

Stretching exercise training as mentioned in below for the period of eight weeks. Training was given in alternate days (Monday, Wednesday and Friday) in a week upto the duration of 60 to 80 minutes followed by the warm up and ends with the warm down exercises.

Stretching Training Schedule-

Stretching training schedule for eight weeks in alternate days (Monday, Wednesday and Friday).

This begins with six to eight minutes of jogging, followed by neck, shoulder, lower back and abdominal stretches. There should be two to three different routines, with 5-10 repetitions of each.

Next, the legs (hamstrings, hip flexors, abductors, abductors, quads and calf muscles) are targeted, with passive (held) and dynamic stretches (two to three standard routines with 10-12 reps, with performance speed increased every set of the dynamic stretches, such as leg swings). Each stretch should be held for 20-30 seconds. And few stretches should be held for 6-10, 10-15.

- Do not hold a stretch that is in anyway painful. It should feel tight and that tightness should diminish as you hold the stretch.
- Breath. Avoid breathing holding while doing stretch as this can raise blood pressure and leave us feeling dizzy.
- Makes sure body is in completely warm before the start. Either do 5-10 mins of light warm up exercise.

Schedule-

- Training was performed regularly in evening session (4.30pm to 5.30pm).
- Group-I (Stretching Training) were performed the same schedule followed by alternate days.
- Training begin with warm-up and end with warm down.
- The subjects were properly motivated to do their best in training which they undergo for eight weeks. Because of their full co-operation, the study was completed in a successful manner.

Table No. 3.2: Stretching Schedule

Exercise	Position	Reps	Holding Sec.
Neck Stretch	Standing	5 – 10	4 Sec.
Triceps Stretch	Standing	5 – 10	6-10 Sec.
Shoulder Stretch	Standing	5 – 10	6-10 Sec.
Chest Stretch	Standing	5 – 10	6-10 Sec.
Sideward & Forward bend	Standing	5 – 10	6-10 Sec.
Hamstring Stretch	Sitting/ Lying	10 – 12	10-15 Sec.
Calf Stretch	Sitting/ Standing	10 – 12	10-15 Sec.
Quadriceps Stretch	Lying / Standing	10 – 12	10-15 Sec.
Groin	Sitting/ Standing	10 – 12	10-15 Sec.

Rest days: Saturday, Sunday, Tuesday and Thursday.

- Stretch the muscles on both sides of the body.
- Do not exceed the threshold for discomfort or pain.
- Do not bounce or jerk, but rather gradually induce the stretch.

Neck Stretch

- Stand with feet hip-width apart, knees slightly bent, hands by your sides (or keep your hands on the hip).
- Turn your head to look over your right shoulder.
- Turn your head to the front and then over to your left shoulder.
- Repeat.
- When you have completed your repetition take your chin down to rest on your chest.
- Bring your chin back up to the centre. Never take your head backwards.
- Repeat.

Shoulder Stretch

- Stand with feet shoulder width apart, knees slightly bent. Place your right arm across your chest and bring your left arm up, bent to hold your right elbow against your chest.
- Pull your left forearm towards your body, pressing on the right elbow and hold for 6-10 seconds.
- Swap arms and repeat.

Chest Stretch

- Stand with hip width apart, knees slightly bent and hands clasped together behind you back (do not interlock fingers).
- Raise your arms as high as you can behind you and hold for 6 – 10 seconds.

Triceps Stretch-

- Stand erect, place your left hand behind your head and reach as far down your back as possible.
- With your right hand your left elbow and gently pull it behind the back of your head.
- You can perform this exercise sitting or standing position. Repeat for the other arm.
- Hold for 6 -10 sec.
- Target muscle: Triceps, deltoid.

Calf muscle stretch-

- Stand erect position, then take one leg forward. (One step) and the heal touch on the ground.
- Body bent forward and both hands grasp the toes.
- The other leg should be slightly bent.
- The head should be downward.
- Hold 10-12 seconds repeat for the other leg.
- Target muscle: gastrocnemius.

Hamstring stretch

- Make sitting Position, and one leg should be extend straight other leg should be in side, bend the feet touch on the other thigh.
- Both hands to touch the toe of extending leg. This exercise perform bending position.
- Hold 10-12 sec and repeat for the other leg.
- Target muscle: hamstring.

Standing Calf Stretch

- Using a wall or bar to support you, one leg out stretched behind you,

- Keeping the other leg forward , lean against wall to apply pressure to your beg leg.
- Make sure you keep your back heel that on the ground.
- Repeat for the opposite leg with 10 – 12 seconds.
- Target muscle: gastrocnemius.

Groin Stretch

- Stand with your legs wider than shoulder with apart. Shift your weight onto one side as you bend your knee.
- Reach with one hand towards your out stretched foot.
- You should feel the stretch right down the inside of your out stretched leg.
- Repeat for the opposite side.
- Target Muscle: groin.

Groin Stretch (Sitting position)

- Sit with the soles of your feet together and hold on the your legs just above the ankles, with elbows placed on the inside of your knees.
- Use your elbows to push your knees from towards the floor.
- Hold for 20 – 30 seconds.
- Target muscle: groin

Side Stretch

- Stand with feet 18-24 inches apart, knees slightly bent. Place your left hand on your thigh and reach up with your right arm.
- Lean over to the left and hold for 6-10 second. Do not lean forward or backwards and do not lock the knees.
- Swap arms and repeat to the other side.

Forward bend

- Stand with feet hip-width apart knees slightly bent, hands on hips.
- Bend forward from the hips and then back up to the vertical position.
- Do not lean backward when coming back up.
- Repeat.

The Physical Fitness variables data was to be elicited by conducting specific activities such as race and other exercises. The specific tools administered and units of measurement are presented in table 3.2.

Table No. 3.3: Selected Physical Fitness Variables, Tests and Criterion Measures

S.N.	Variables	Tests/Tools Administered	Unit of measurement
1	Arm Power	Two hand medicine ball throw	In Meters
2	Leg Power	Standing broad jump	In Meters
3	Speed	50 Mts dash	In Seconds
4	Agility	4 X 10 Mts shuttle run	In Seconds
5	Abdominal Endurance	Bent Knee Sit ups	Number of repetition in 60 Sec

3.7 Criterion Measures-

Stretching-

Self-made questionnaire was used analyse the Stretching Effectiveness of the subject and subjects response converted into numbers by using strategy.

In Stretching Effectiveness questionnaire there 65 questions were made by the researcher and send them on the expert of this field. The expert gives his suggestion on the questionnaire and cut duplicate and non-related question in the questionnaire. Finally 40 questions were selected for the questionnaire.

Physical Fitness-

1. Two hand medicine ball throw test was employ to measure arm power of the subject and the distance was measured to the nearest centimeter.
2. Standing broad jump test was used to measure leg power of the subject and the distance recorded to the nearest centimeter.
3. 50 meters dash test was used to calculate speed of the subject and time was recorded in 1/100th of a second.
4. 4 x 10 meters shuttle run test was employ to measure agility of the subject and time was recorded in 1/100th of a second.
5. Bent knee sit-ups test was used to measure abdominal endurance of the subject and the score was taken completed sit-ups in 60 seconds.

3.8 Administration of Tests-

To test stretching and physical fitness were conducted where adequate facilities are available like stadium, college ground and class room's facilities are used to collect the data. The objectives and importance of the study was told to the players prior to the measurements to motivate them and to get their cooperation during the tests. The support of physical education directors, team managers and coaches was taken to contact the players. The physical education, managers and coaches of the respective teams ensured the availability of players for the collection of data. Demonstration of all the tests was given before the players and all kinds of efforts had been made by the researcher to make sure accuracy and uniformity within the administration of the test. Before conducting physical fitness test a short warming up period of 10 mints duration was given to all the subjects. All the tests were conducted on each subject.

Physical Fitness-

1. Arm Power: Two Hand Medicine Ball Throw

Objective: To measure the shoulder and arm power of the subject

Equipment's: Chair, flexible measuring tape, marking powder, and 6 pound medicine ball is required conduct the test.

Procedure: From a sitting position in a straight back chair, the performer holds the ball in both hands with the ball drawn back against the chest and just under the chin. He then pushes the ball upward and outward for maximum distance. The rope is placed around the performer's chest and held taut to the rear by a partner in order to eliminate rocking action during the push. The performer's effort should be primarily with the arms.

Scoring: The distance of the best of three trials measured to the nearest foot is recorded is measured from the forward edge of the chair to the point of contact of the ball with the floor.

2. Leg power: Standing Broad jump

Objective: To measure the Leg power of the subject.

Equipment and marking: Measuring tape, long jump landing pit and marking powder are required to conduct the test.

Procedure: The subject stood the takeoff line with feet parallel to each other. The performer flexed his knees and took his arms backward then with a vigorous forward swing of arms and extension of flexed knees he took off in one chance and jumped on the lading pit as for forward as possible. These trials were given after adequate rest.

Scoring: Best of the three trials in measure was considered as the best performance of the subject in the tests.

3. Speed: 50 Meter Dash

Purpose: To measure the running speed of the subject.

Equipment's: stopwatch.

Procedure: The purpose of 50 m dash trial is to examine the acceleration and speed of the subject. The subjects were instructed to stand in stationary position at the back of the starting line and the tester gave the command to subject set and then go. The subject took start from a stationary standing position and had to cross the finish line with maximum possible speed.

Scoring: The time taken to complete the 50 m distance was recorded to the nearest 1/100th of a second (Johnson and Nelson, 1979).

4. Agility test: 4 x 10 meters shuttle Run

Purpose: To assess the agility of the subject.

Procedure: To conduct the test, at ten meters of distance two parallel lines of five meters width will be marked. The subject was asked to stand behind the starting line. On getting the starting signal "Go", the subject run faster, goes nearer to the other line and touches it with the one hand, turns and came back to starting line and touches it by hand. The subjects were asked to do the shuttle run four times and finish across the starting line by run through.

Scoring: The time elapsed between start signal "go" and four shuttle runs finishing has been recorded to nearest 1/100th of a second. Two trials were given to the subject. The best performance was used as a score for the purpose of the study (Mathews, 1973).

5. Abdomen Endurance: Bent Knee Sit-ups

Objective: To measure the abdomen endurance of the subject.

Equipment and marking: Stop watch and mats otherwise floor is satisfactory.

Procedure: The subject lies flat on the back with knee bent and feet on the floor with the heels no more than one foot from the buttocks. The knee angle should be no less than 90 degrees. The fingers are interred locked and placed behind the neck with the elbows touching floor. The feet are held securely by a partner, when ready the tester give start signal, the subject then curl up a sitting position and touches the elbows to the knees. When he returns to the starting position, his elbows must be flat on the floor or mat. He performs sit up as fast as he can for 60 sec. The tester gives stop signal after 60 seconds.

Scoring: The completed sit-ups in 60 seconds are taken (Barrow and McGee, 1979).

3.9 Stretching Effectiveness Questionnaire-

- **Purpose-**

The self-made Effectiveness of Stretching on Performance of Softball Players. The Purpose of the test is to find out the stretching effectiveness capacity.

- **Tool-**

The self-made tool constructed by researcher in guidance of supervisor.

- **Procedure-**

The researcher himself will administer the stretching questionnaire to the players through personal contact at the time of inter collegiate Softball tournaments. The questionnaire has 40 items. Ever player has to answer all 40 items of the tests and was instructed to express his choice. The choice was selected in five points scale of questionnaire that were **Strongly Agree, Agree, Uncertain, Disagree, Strongly Disagree.** The method of answering tick mark (√) was explained to them.

- **Scoring-**

Subject response converted into numbers by using answer keys. In which **Strongly Agree** was marked by **'5', Agree** was marked by **'4', Uncertain** was marked by **'3', Disagree** was marked by **'2' and Strongly Disagree** was marked by **'1'.** The marks of questionnaire were decided in the range of 40 to 200 marks.

- **Reliability**

The coefficient of reliability was determined by using the product moment correlation was 0.85 and by using Spearman Brown formula was 0.92. Both the values ensure a high reliability of the test.

- **Validity**

The coefficient of validity was 0.62, which is significant beyond 0.01 level of confidence.

- **Time**

There is no time limit to complete the test. Generally, 15 to 20 minutes are sufficient to complete the test.

- **Collection of data-**

The data will be collected with self-made stretching effectiveness questionnaire. The data will be collected from Softball players who were participated in the Inter Collegiate and inter-zone softball championship. Researcher collects data from various colleges for raw score.

- **Procedure of the Study-**

The researcher initially tried to contact to organizing authorities of tournament and received permission to collect the data. Convenient meeting dates, time and place arranged with Softball players. Before filling the questionnaire instructions printed on the questionnaire were explained verbally so that the subjects do not hesitate to give their honest, personal and frank response in the correct manner. Before the filling actual questionnaires, each participant completed a background information such as age, participation in game, level of participation, Height, Weight and Group. This information was used to determine Effectiveness of Stretching on performance of Softball Players of the participants. The questionnaires were responded in a quiet, comfortable, and familiar environment at the participant's living rooms and grounds.

3.10 Assessment of Performance of Softball-

- **Players-**

After consulting with experts in Softball, a performance of rating scale was developed by the Investigator to assess the performance of Softball players in different areas of the game.

- **Performance Rating Scale-**

The Rating Scale was developed according to the opinion and suggestions of the experts in the field. The following persons who are in the coaching field in the Softball game currently, Coach of Indian Softball team and softball expert were consulted. After a thorough discussion among the experts, the guide and research scholar, the performance rating scale of softball players was formulated. The experts suggested combining both subjective and objective types of rating scales to assess total performance of the softball players.

Regarding this study, the data pertaining to the performance ability inter collegiate softball players are gathered by adopting rating method. A panel of three expert coaches rates the subject's performance in various parameters like skill, technique and application of skill in the game situation, such as batting, throwing, catching, stealing ability, application of techniques, tactics, strategies and Fielding ability are assessed on 10 point rating scale, the rating will be based on subjective evaluation to predict the performance ability of softball players.

Instructions for the Expert / Coach to Evaluate the Performance Ability of the Players:

1. The performance ability of male softball players was examined by the experts/coaches in the competitions on a ten (10) point rating scale.
2. The rating scale at ten categories with each category being scored a minimum of one (1) point to a maximum of ten (10) points.

3. The experts were informed to follow a uniform pattern of game observation, guidelines and performance chart.
4. Experts' coaches inform to encircle the number in each category.
5. The experts/coaches will inform to assess each category separately and independently.
6. Points scored by each player in each category have to be added to get the performance rating of each individual player. The scores / points awarded by Experts / coaches on the scale were totalled. Higher scores would denote better performance.

The experts/coaches were orientated by the researcher regarding the above procedures, to be taken into consideration during the process of evaluating the game performance ability of the male softball players. Thus the performance scores for the softball players will be obtain.

Table 3.4: Performance Rating Scale for Assessment of Performance Ability of Softball Players

Game Performance				
Sl. No	**Variables**	**Performance**	**Measures in Points**	**Scored Points**
1	Hitting Ability	**1 2 3 4 5 6 7 8 9 10**	**10**	
2	Throwing Ability	**1 2 3 4 5 6 7 8 9 10**	**10**	
3	Base Stealing Ability	**1 2 3 4 5 6 7 8 9 10**	**10**	
4	Fielding Ability	**1 2 3 4 5 6 7 8 9 10**	**10**	
5	Catching Ability	**1 2 3 4 5 6 7 8 9 10**	**10**	
6	Base Running Ability	**1 2 3 4 5 6 7 8 9 10**	**10**	
7	Game Tactics	**1 2 3 4 5 6 7 8 9 10**	**10**	
8	Overall Game Performance	**1 2 3 4 5 6 7 8 9 10**	**10**	
9	Past and Present Achievement	**1 2 3 4 5 6 7 8 9 10**	**10**	
10	General Behavior	**1 2 3 4 5 6 7 8 9 10**	**10**	
		Total Point Scored by Player	**100**	

3.11 Statistical Procedure-

Completion of any scientific analysis is possible only with the use of statistical processing. The acceptance or rejection of hypothesis ultimately determines the contribution of the investigation in the scientific development of a particular area. This is especially true for statistical techniques in the analysis for interpretation of the data with the use of some statistical processing.

The objective of the study was to compare the data of Stretching Effectiveness and physical fitness with the performance of the softball players. The data analyzed as descriptive statistics such as mean, standard deviation, standard error of mean, minimum value and maximum value. The scores of the softball performance were converted into standardized scores. Karl Pearson's product moment co-efficient of correlation was computed to assess the relationship of Stretching Effectiveness and physical fitness variables with the performance of the softball players. To predict the performance on playing ability in softball from Stretching Effectiveness and physical fitness parameters, stepwise regression analyses were applied.

In this study researcher used four popular methods which are –

1. Mean
2. Standard Deviation
3. T-test

Researcher has selected two levels of significance as 0.05 and 0.01. It is defined as for the degree of improbability which is deemed necessary to cast sufficient doubt upon the truth of the hypothesis to warrant its rejection.

3.11.1 Statistical Procedure-

In order to find self-concept, achievement motivation and adjustment of the basketball players of different levels of competitions, data were summarized by descriptive statistics (mean, standard deviation).

Mean -

The statistical mean refers to the mean or average that is used to derive the central tendency of the data in question. It is determined by adding all the data points in a population and then dividing the total by the number of points. The resulting number is known as the mean or the average. In mathematics and statistics, the term arithmetic mean is preferred over simply "mean" because it helps to differentiate between other means such as geometric and harmonic mean. Statistical mean is the most common term for calculating the mean of a statistical distribution.

M= A.M. + Σfd X C.I. /N

A.M.= Assumed mean

d = X-A/C

C.I.= Class Interval

N = Total Number of Sample

Standard Deviation -

The best measure of dispersion is, usually, standard deviation which does not possess the demerits of range and mean deviation. Standard deviation for a given set of observations is defined as the root mean square deviation when the deviations are taken from the AM of the observations.

S.D. = C.I.√NΣfd2 – (Σfd)2/N

d = X-A/C

C.I.= Class Interval

N = Total Number of Sample

't'- value -

A t-test is any statistical hypothesis test in which the test statistics follows a Student's t distribution if the null hypothesis is supported. It can be used to determine if two sets of data are significantly different from each other, and is most commonly applied when the test statistic would follow a normal distribution if the value of a scaling term in the test statistic were known. When the scaling term is unknown and is replaced by an estimate based on the data, the test statistic (under certain conditions) follows a Student's t distribution.

$t = M1 - M2 / \sqrt{(\sigma1)2/N1 + (\sigma2)2/N2}$

M1 = Mean of first sample

M2 = Mean of second sample

σ1 = Standard deviation of first sample

σ2 = Standard deviation of second sample

N1 = Total number of first sample

N2 = Total number of second sample.

Chapter IV

ANALYSIS OF DATA AND RESULT OF THE STUDY

4.1 Introduction -

The present chapter deals with the analysis and interrelation of the obtained data with reference to the objectives and hypothesis. A prime responsibility of the educational researcher is of being able to make either a probability or logical influence covering the tenability of his testable hypothesis what contribution the study makes to the scientific development of a particular area. This is especially tried out in the analysis and interpretation of data.

"The mountain of data does not say anything unless and until it is given an organized form. Therefore, the classification and tabulation of data is significant."

Analysis of data includes comparison of the outcomes of the various treatments to achieve the goals of research. Data relevant to each hypothesis can be accepted or rejected. The present study was aim to analysed the data and interpret the result regarding Effectiveness of Stretching on Performance of Softball Players.

4.2 Analysis and Interpretation of Data-

Analysis-

According to Goods Barr and Scates-

"Analysis is a process which enters into research is one form or another form in very beginning… it may be fair to say that research contains in general of two larger steps the gathering of data and the analysis of data, but no amount of analysis can validly exact from the data factors which are not present."

Analysis of data included comparison of the outcome of the various treatment upon the several group and the making of a decision as to the achievement of the goals of research data relevant to each hypothesis, must be assembled in quantitative form and tested to determine whether or not the hypothesis can be accepted or rejected.

Analysis means categorizing ordering, manipulating and summarizing data to obtain answer to research questions. The purpose and summarizing data to obtain answer to research question. The purpose of analysis is to reduce data to intelligence and interpretable form so that the relation of research problems can be studies and tested. The research should lay out analysis paradigm of mode is even while working on the problems and hypothesis. Only in the way we can see it deeply whether in data and its analysis will answer the research problems.

Interpretation-

Interpretation takes into account the results of analysis and makes the inferences pertinent to the research problem under investigation and draws generalization about these problems. The investigator who interprets results, researches these for their meaning and implications. This can be done in two ways:-

1. Narrower way in which the relation with the research study and its data are interpreted. In this type interpretation and analysis are clearly interviewed.

2. Broader way in which the results and inferences drawn within the data are compared to the theory and other research results and conclusions either to one's own or the other researches.

The research is incomplete without the analysis and interpretation of data that's why they are regarded as a part of research. The data are often classified by division into sub groups and are then analyzed and synthesized in such a way that hypothesis may be verified or rejected. The final result may be a new principle or a generalization.

"The discovery of order in the phenomena of nature not withstanding their complicity and apparent confusion in renderer possible by the process of analysis and synthesis which are the foundation stones of all scientific methods." **-Wolfe**

The mass of data collected through the use of various tools, however reliable, valid and adequate it may be is yet but raw. It needs to be systematized and organized i.e. edited, classified and tabulated before it can serve any worthwhile purpose.

4.3 Process of Analysis of Data -

The analysis of data requires a number of closely related operations; such as establishment of categories of raw data through coding tabulation and then drawing statistical inferences. In the process of analysis, relationship or differences, supporting or conflicting with original and new hypothesis should be subjected to test of significance to determines with original and new hypothesis should be subjected to test of significance to determine with what validity data can be indicate any conclusions through the use of statistical tests. We can establish whether such a difference is real or is the result of random fluctuations.

"Analysis means the categorizing manipulating and summarizing of data to obtain answer to research questions." **-Kerlinger**

The purpose of analysis is to reduce data to intelligible and interpretable form so that the relation between various elements of data with reference to research problem can be studied and tested. **Good, Barr and Scates** suggested four helpful modes to start with the analysis of data:

1. To think in terms of significant tabled that the data permits.
2. To examine carefully the statement of the problem and the earlier analysis to study the original records of data.
3. To get away from the data top think about the problem in Lay Men's terms or to actually discuss that problem with others.

4.4 Tabulation of Data -

Tabulating denotes the recording of the classified in accurate mathematical terms e.g. making and counting frequencies tallies for different items on which information is gathered. It is a demanding process and it must be accurate.

But tabulation is possible to clarity the clear picture of data because in it, data are systematically arranged, which makes comparison of data possible. Tabulation proves helpful in saving of time and place. A good table must be attractive and of appropriate size Table should be according to objectives to study and must be clear and built by scientific method.

4.5 Analysis and Interpretation of Data of Current Study -

The analysis and interpretation of the collected data is the most important and inseparable part of any research work to analyze the data and give the appropriate interpretation for the same.

In the following pages of the chapter the researcher has tried to make an attempt to accomplish this work keeping in mind the importance of the work in the current research.

Results have been presented in the following tables:

Table 4.1– Table showing significant difference of effectiveness of stretching on performance of the Softball Players.

Table - 4.1

Source of Variation	N	Mean	S.D.	Df.	t-value	Level of significance
Boys	85	162.20	12.07	168	2.67	S*
Girls	85	167.46	13.54			

***Significant at 0.01 & 0.05 level**

Table 4.1 significant difference of effectiveness of stretching on performance of the Softball Players -

Description – The table 4.1 shows that Mean score of boys and girls softball players are 162.20 and 167.46 with a **SD** values of boys and girls softball players are 12.07 and 13.54 respectively. t-value **(2.67)** is found to be significant at **(0.01 & 0.05)** level. Hence the null hypothesis is rejected which means that **there is significant difference of effectiveness of stretching on performance of the Softball Players.**

The results obtained from this table 4.1 are cleared from following graph –

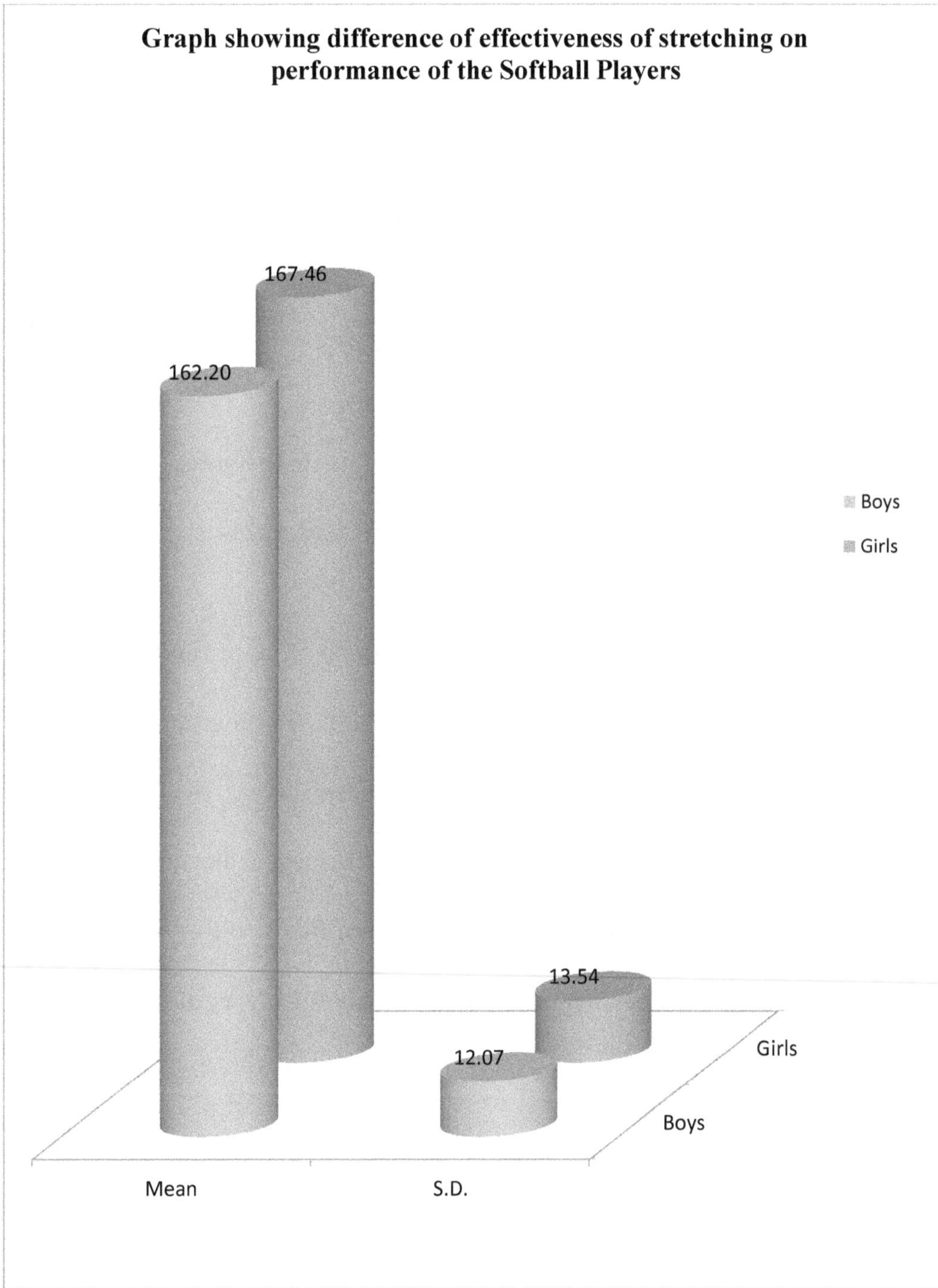

Table 4.2– Table showing significant difference of effectiveness of stretching on performance of the Softball Players of Rural Area.

Table - 4.2

Source of Variation	N	Mean	S.D.	Df.	t-value	Level of significance
Rural Boys	40	165.30	11.66	78	0.62	NS*
Rural Girls	40	167.10	14.12			

***Not Significant at 0.01 & 0.05 level**

Table 4.2 significant difference of effectiveness of stretching on performance of the Softball Players of Rural Area-

Description – The table 4.2 shows that Mean score of rural boys and rural girls softball players are 165.30 and 167.10 with a **SD** values of rural boys and rural girls softball players are 11.66 and 14.12 respectively. t-value **(0.62)** is found to be not significant at **(0.01 & 0.05)** level. Hence the null hypothesis is not rejected which means that **there is no significant difference of effectiveness of stretching on performance of the Softball Players of Rural Area.**

The results obtained from this table 4.2 are cleared from following graph -

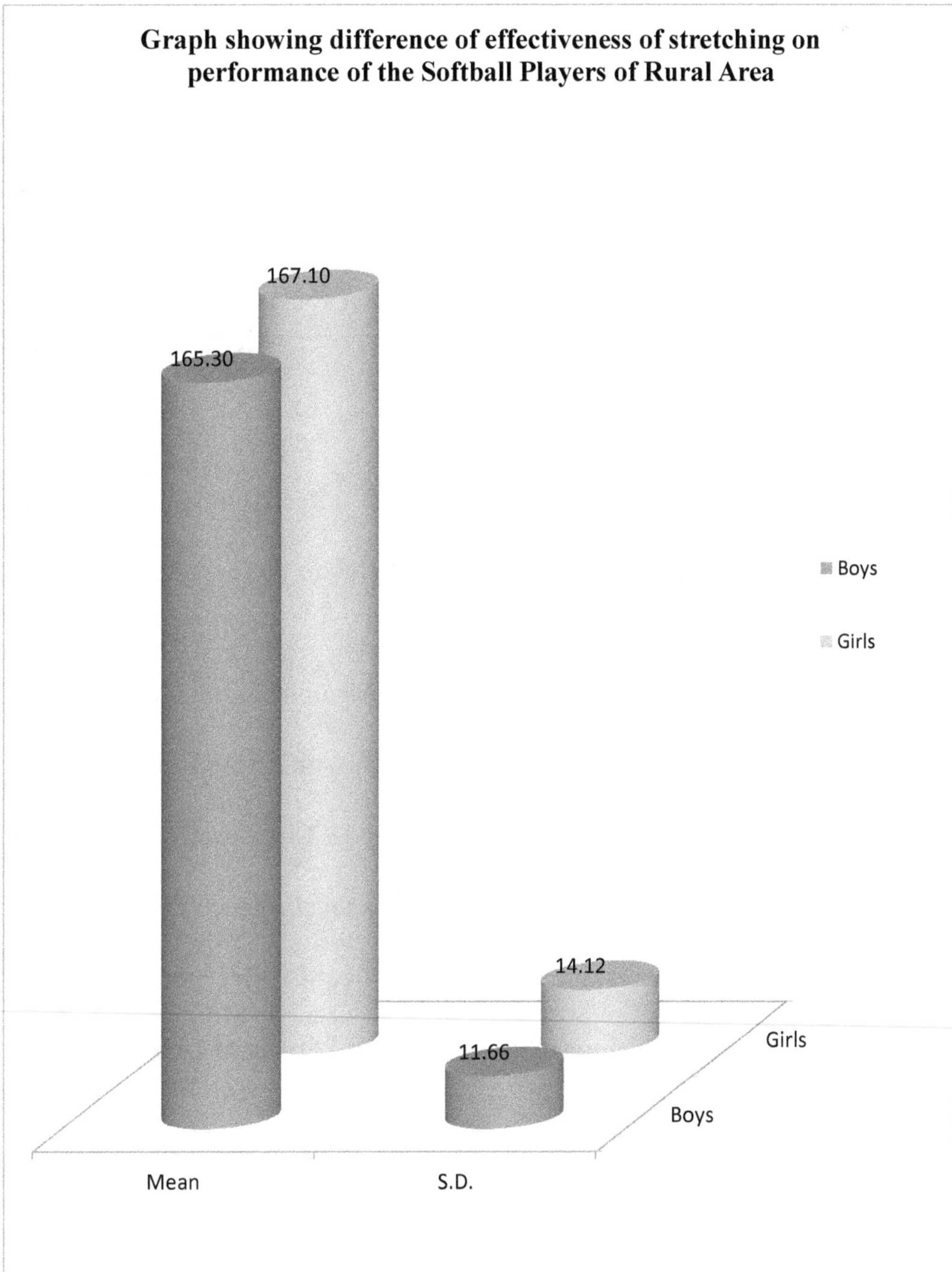

Table 4.3– Table showing significant difference of effectiveness of stretching on performance of the Softball Players of Urban Area.

Table - 4.3

Source of Variation	**N**	**Mean**	**S.D.**	**Df.**	**t-value**	**Level of significance**
Urban Boys	45	159.44	11.87	88	3.16	S*
Urban Girls	45	167.78	13.15			

***Significant at 0.01 & 0.05 level**

Table 4.3 significant difference of effectiveness of stretching on performance of the Softball Players of Urban Area-

Description – The table 4.3 shows that Mean score of urban boys and urban girls softball players are 159.44 and 167.78 with a **SD** values of urban boys and urban girls softball players are 11.87 and 13.15 respectively. t-value **(3.16)** is found to be significant at **(0.01 & 0.05)** level. Hence the null hypothesis is rejected which means that **there is significant difference of effectiveness of stretching on performance of the Softball Players of Urban Area.**

The results obtained from this table 4.3 are cleared from following graph -

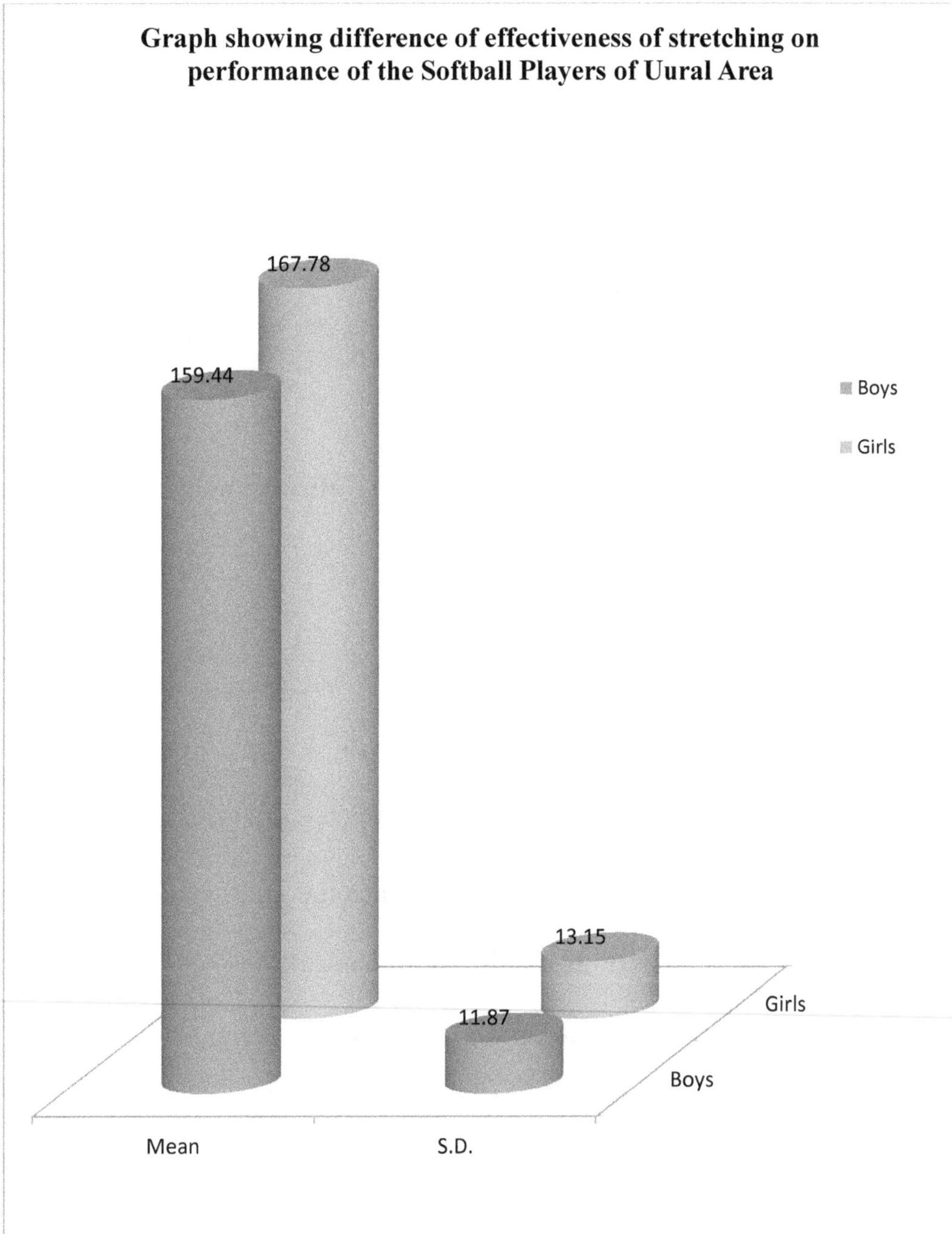

Table 4.4– Table showing significant difference of effectiveness of stretching on performance of boys' Softball Players.

Table - 4.4

Source of Variation	N	Mean	S.D.	Df.	t-value	Level of significance
Rural Boys	40	165.30	11.66	83	2.29	NS*
Urban Boys	45	159.44	11.87			

***Not Significant at 0.01 levels**

Table 4.4 significant difference of effectiveness of stretching on performance of boys' Softball Players -

Description – The table 4.4 shows that Mean score of rural and urban boys' softball players are 165.30 and 159.44 with a **SD** values of rural and urban boys' softball players are 11.66 and 11.87 respectively. t-value **(2.29)** is found to be not significant at **(0.01)** level. Hence the null hypothesis is not rejected which means that **there is no significant difference of effectiveness of stretching on performance of boys' Softball Players.**

The results obtained from this table 4.4 are cleared from following graph –

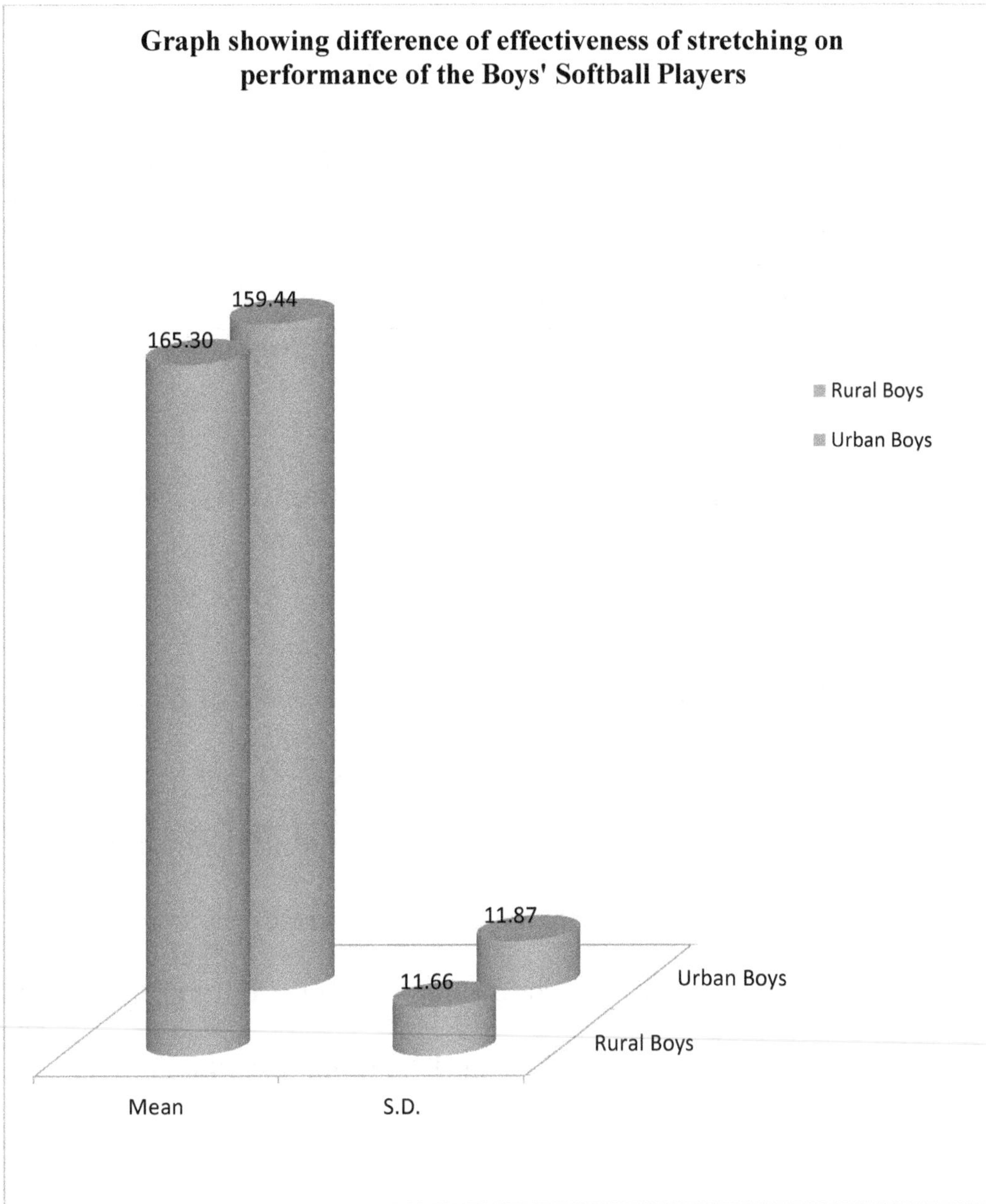

Table 4.5– Table showing significant difference of effectiveness of stretching on performance of girls' Softball Players.

Table - 4.5

Source of Variation	N	Mean	S.D.	Df.	t-value	Level of significance
Rural Girls	40	167.10	14.12	83	0.23	NS*
Urban Girls	45	167.78	13.15			

***Not Significant at 0.01 & 0.05 level**

Table 4.5 significant difference of effectiveness of stretching on performance of girls' Softball Players -

Description – The table 4.5 shows that Mean score of rural and urban girls' softball players are 167.10 and 167.78 with a **SD** values of rural and urban girls' softball players are 14.12 and 13.15 respectively. t-value **(0.23)** is found to be not significant at **(0.01 & 0.05)** level. Hence the null hypothesis is not rejected which means that **there is no significant difference of effectiveness of stretching on performance of girls' Softball Players.**

The results obtained from this table 4.5 are cleared from following graph –

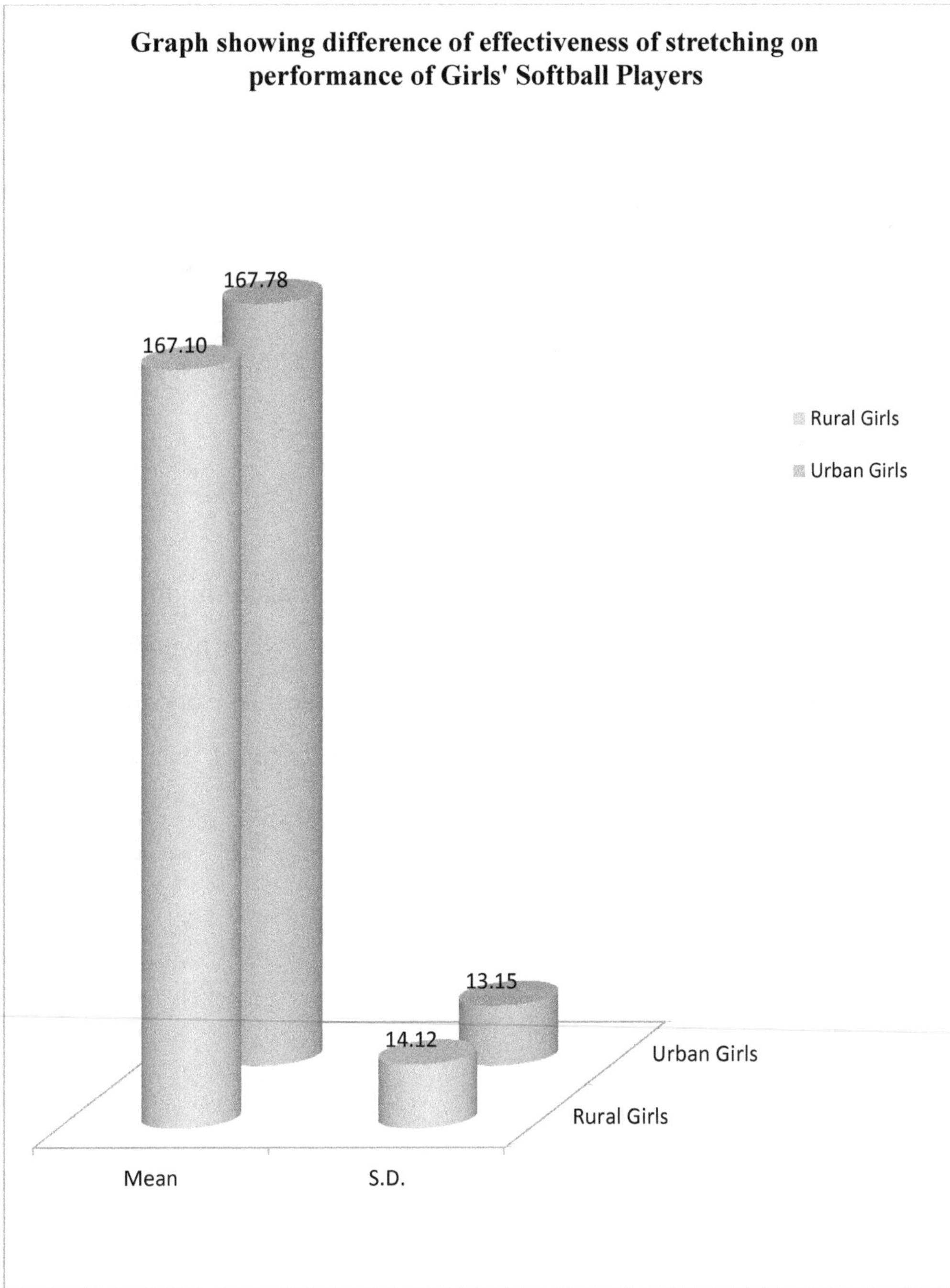

4.6 Result of Hypothesis -

1. There is significant difference of effectiveness of stretching on performance of the Softball Players.
2. There is no significant difference of effectiveness of stretching on performance of the Softball Players of Rural Area.
3. There is significant difference of effectiveness of stretching on performance of the Softball Players of Urban Area.
4. There is no significant difference of effectiveness of stretching on performance of boys' Softball Players.
5. There is no significant difference of effectiveness of stretching on performance of girls' Softball Players.

4.7 Conclusion -

The current research work was undertaken in the effectiveness of stretching on performance of softball players of some hypothesis. Therefore it now becomes essential at this stage of the research work to see whether the hypothesis were rejected or accepted on the basis of data analyzed. In next chapter we summarized the research work.

Chapter-5

Summery, Conclusion & Recommendation

5.1 Introduction -

Games and sports have been part of human life since the ancient days. It is a basic requirement for survival i.e. hunting for food, sanctuary and safety from wild animals or other enemy or as a hunt of pleasure. The games and sports have been indispensable to mankind and have become part of his culture also. The games and sports are a great unifying force and have tremendous effect on the national and international integration.

Sport is the means in which we use our physical capacities to play. Sports is crucial in different ways, while one's body works better his mentality works higher, his brain and his body are interrelated. Sports permit us to blow off the anxiety to overlook problems for some time and to go out and have a terrific time regardless of what other pressures one can be beneath in his lifestyles. Sports are in man's blood, Sport is recreation as well as competition. Sports are essentially person activities referring to and revitalizing in nature and meant to provide opportunities to the man or woman to make two fullest and maximum smart use of amusement time.

Sports and games are a way of maintaining physical fitness, apart from promoting recreation and showing one's upper-handiness over the other. Excelling the tactics of sports events, many legends have emerged in which led to its evaluation and development of different events. Throwing some light on the reasons and circumstances would help us in better understanding the field of sports.

Struna (1977), rightly point out that the sport has emerged under certain social context and it use understanding which becomes important to appreciate the true spirit of sports, again finds that there are multiple contexts in which the evolution can be explained based on the post experiences. Such exploration led to the study of a range of sports topics such as Physical Education, dance and play.

5.2 Softball -

Softball is a baseball descendant game that requires speed, strength, and endurance. Softball is a bat-and-ball sport that entails skills such as throwing, fielding, pitching, catching, base-running and hitting (Craig et al., 1985). To develop in these areas, many factors play a significant role like skill, practice, level of physical activity, techniques, psychological traits, nutrition and other environmental factors. However, a particular body size and shape and certain genetically conditioned abilities and features are required in order to achieve top level performance in softball. The increased competitiveness and the drive to excel in sports and to achieve national or international status in sports demands that players accomplish their optimum performance levels in a systematic way with clearly defined performance objectives and training programs. This has motivated coaches, managers, sports administrators, trainers, sports scientists and players to spend more time and resources assessing their sports performance more scientifically. This requires the assessment of anthropometric characteristics, physiological variables, and sports specific skills for adequate sport performance. Since performance in sports is a multivariate phenomenon, it may be influenced by physique in addition to physiological function, biochemical constraints, psychological state, environment and socio-cultural factors.

Softball is a physically demanding sport comprised of several specializations such as hitting, fielding, throwing, pitching, base running etc. requiring diverse skills and different types of fitness. It requires upper extremity power, hand grip strength, excellent eye-to-hand coordination and the coordinated movements of the hips, shoulders, arms and wrists. It requires speed, strength and endurance (Terbizan et al., 1996). Softball players need a considerable amount of upper body muscle balance because of the specificity of underarm pitching and overhead throwing activities. Coaches and athletes consider that the forearm plays an important part in hitting, pitching and throwing the ball.

In softball, there is little scientific information to suggest which specific components of strength, speed or power best predict softball playing ability. There is also very scant information regarding the anthropometric characteristics as the predictors of performance in softball. As such, the development of performance prediction models as a function of specific tests of physical capacity and anthropometric characteristics may provide insight into those factors which relate to softball playing ability.

Playing softball every day, and sometimes twice a day, takes a physically and mentally fit athlete. The ability to accept failure since it is an inevitable part of the game, while being prepared to perform in the moment that demands excellence takes a special set of skills physically and mentally. From pitching to fielding, and hitting to base running, the softball player must take care of their body through a properly implemented strength and conditioning program that understands the value of recovery, nutrition, and sleep.

The game of softball originated as a kind of baseball for youngsters. Several popular sports had a significant impact on the development of softball, including boating, boxing, and football. There was also a rowing club, boxing glove, and football club involved in the original game of softball.

5.2.1 Playing Method -

Softball is played between two teams on a large field, with 9 players from one team on the field at a time. The field is usually composed of a dirt or brick-dust infield that contains the quadrilateral shape and running areas, and a grass outfield. However, the field can consist of other solid and dry surfaces such as artificial turf or asphalt. There are 4 bases on the infield (first base, second base, third base, and home plate); the bases are arranged in a square and are typically 45 to 65 feet (13,7 to 19,8 meters) apart. Near the center of this square is the pitcher's circle, and within the circle is the "rubber", a small flat rectangular piece of rubber about a foot and a half in length. The rubber can be 40 or 43 feet away from home plate, depending on age level and the league one is playing in. The object of the game is to score more runs (points) than the other team by batting (hitting) a ball into play and running around the bases, touching each one in succession. The ball is a sphere of light material, covered with leather or synthetic material. It is 10 to 12 inches (or rarely, 16 inches) (28 to 30.5 centimeters) in circumference. The game is officiated by one or more neutral umpires. Players and umpires are generally free to ask for a brief stoppage at any time when the ball is not in play, or immediately following a play once its outcome is clear (Austen, Jacob, 2005).

The game is played in usually 7 innings. Each inning is divided into a top half, in which the visiting team bats and tries to score runs, while the home team occupies the field and tries to record 3 outs; then a bottom half, when the teams' roles are reversed. Youth leagues may play 6-inning games; most ASA and middle school rules specify 5 innings and a maximum time of 1.5 hours.

When a ball is batted into play, runners generally must attempt to advance, if there are no open bases behind them; for example, a runner on first base must run to second base if the batter puts the ball in play. In such a situation, the defence can throw to the base that the lead runner is attempting to take (a force out), and the defence can then also throw to the previous base. This can result in a multiple-out play: a double play is two outs, while a triple play, a very rare occurrence, is three outs. Runners with an open base behind them are not forced to advance and do so at their own risk; the defence must tag such runners directly to put them out rather than tagging the base. A ball hit in the air and caught before hitting the ground, in fair or foul territory, puts the batter out. A fly ball is a ball hit high and deep, a pop fly is a ball hit high but short, and a line drive is a ball hit close to the horizontal. After the catch, runners must return to their original bases; if the defence throws the ball to that base before the runner returns, the runner is out as well, resulting in a double play. A runner who remains on the base until the ball is touched, or returns to the base (tags up) after the catch, may try to advance to the next base, at the risk of being tagged out between bases. As in baseball, the infield fly rule applies in some game situations to prevent the defence from recording multiple force-outs by deliberately dropping an easy catch.

5.2.2 Effectiveness -

Stretching has been found both effective and ineffective based on its application for treatment.

Although many people engage in stretching before or after exercise, the medical evidence has shown this has no meaningful benefit in preventing specifically muscle soreness. It may reduce the lactic acid build up in the muscles, making the next workout more bearable.

Stretching does not appear to reduce the risk of injury during exercises, except perhaps a dynamic warm-up for runners. While running places extreme stress loads on the joints, static stretching can help to improve joint flexibility. However, this has not been proven to reduce risk of injury in the runners. A dynamic (stretching) warm up has been shown to help overall running performance.

5.3 Stretching -

Stretching is a physical exercise that requires putting a body part in a certain position that'll serve in the lengthening and elongation of the muscle or muscle group and thus enhance its flexibility and elasticity.

The stretching of a muscle fiber begins with the sarcomere, the basic unit of contraction in the muscle fiber. As the sarcomere contracts, the area of overlap between the thick and thin my filaments increases. As it stretches, this area of overlap decreases, allowing the muscle fiber to elongate. Once the muscle fiber is at its maximum resting length (all the sarcomeres are fully stretched), additional stretching places force on the surrounding connective tissue. As the tension increases, the collagen fibers in the connective tissue align themselves along the same line of force as the tension. Therefore when you stretch, the muscle fiber is pulled out to its full length sarcomere by sarcomere, and then the connective tissue takes up the remaining slack. When this occurs, it helps to realign any disorganized fibers in the direction of the tension. This realignment is what helps in the rehabilitation of scarred tissue.

5.3.1 Effects of Stretching -

- Improves flexibility delaying impaired mobility associated with aging
- Improves performance in physical activities as a result of improved flexibility
- Reduce the risk of injury and damage
- Improves blood circulation reducing muscle soreness and time for recovery
- Increasing the range of motion.

5.3.2 Types of Stretching -

Stretching exercises have traditionally been included as part of the training and recovery program. It's important to note that, maximal strength, number of repetitions and total volume is different for each type of stretching.

The three different types of stretching are:

- Static stretch (SS)
- Dynamic stretch (DS)
- Pre-contraction stretching: Proprioceptive Neuromuscular Facilitation stretching (PNFS).

5.3.3 The Benefits of Softball Stretches -

- Coaches place a Lot of emphasis on together with softball stretches in the numerous softball training packages. This is due to the huge wide variety of advantages of stretching, which includes: Regular softball stretches are key to retaining flexibility, which in flip fosters an awesome posture and reduces decrease back ache and discomfort.
- Stretching sports in softball schooling can increase a participant's athleticism via controlling muscle imbalances, which reason pulled muscle groups and also contribute to clumsiness, which in itself can cause damage.
- Even the most effective softball stretches, as a part of your typical softball education application allows promoting blood movement, which improves intellectual alertness and coordination. Finally, regular softball stretches can assist save you injuries.

5.3.4 Stretching Routine for Softball Players -

No matter what position you play in softball, you'll help prevent injury and ensure your body is warmed up and ready to play by engaging in stretching exercises. Stretching exercises help increase blood flow to all parts of your body and it also increases flexibility and range of motion. Your coach might have specific exercises for you to perform before game time or you can do self-directed stretching exercises.

Rotator Cuff

Warm up your shoulder joint and rotator cuff muscles by doing a rotator cuff stretch before every game. This stretch warms up the entire shoulder and encourages flexibility and full range of motion. To do this stretch, stand with your feet shoulder-width apart. Place your right hand behind your back, with the top of your hand against your spine. Grasp your right elbow with your left hand and gently pull your right elbow forward. Hold the stretch for five seconds and then release. Switch sides and repeat.

Hip Rolls

Hip rolls help loosen the pelvic muscles, stretch the hip joints and increase flexibility in your lower spine. This exercise is done lying on the ground on your back. Bend your knees and stretch your arms out at shoulder height. Slowly drop your knees to the left side of your body, but keep your left hip firmly planted on the ground. Hold the stretch for several seconds and then lift your knees and rotate them to the right. Repeat this back-and-forth roll 10 times.

Dynamic Stretches

Stretches that encourage motion are performed through dynamic stretches. Such stretches are not static but rather encourage movement while warming up muscle groups. You can perform softball dynamic stretches with slow running butt kicks and high knee raises while slow jogging. To perform a butt kick, take small steps forward, lifting first your right leg and then your left behind you high enough for your heels to touch your buttocks. Perform a high knee jog while jogging forward, lifting your knees to the waistline, alternating knees while moving forward.

Shaking Arms

An effective loosening stretch for the hands, arms and shoulders can be accomplished with a shaking arms exercise. Have a partner stand in front of you, holding your hands. Gently yet firmly, your partner shakes your hands up and down. Keep your wrist, elbow and shoulder joints loose during the shaking motion. If you prefer, shake one arm at a time and then switch sides. Shake each arm for approximately 10 to 20 seconds for an effective warm up and stretch.

5.3.5 The Three Best Softball Stretches -

Stretching is one of the most beneath-utilized techniques for enhancing athletic overall performance, stopping sports activities injury and properly rehabilitating sprain and stress damage. Don't make the mistake of questioning that something as easy as stretching won't be effective.

Below are 3 of the excellent stretches for softball; obviously there are plenty more, however those are a first rate area to start. Please make unique note of the commands with each stretch, and if you presently have any persistent or habitual muscle or joint pain please take extra care when appearing the stretches under, or seek advice from your doctor or physical therapist before appearing any of the following stretches.

Lying Knee Roll-over Lower Back Stretch: While Lying on your returned, bend your knees and allow them to fall to 1 aspect. Keep your fingers out to the facet and permit your again and hips rotate with your knees.

The lying knee roll-over lower back stretch explores loosening up restrictions in both the lumbar spine (lower back) and thoracic spine (mid-back). An inability to rotate the spine effectively will impact your ability to throw and hit well in softball, so including the lying knee roll-over lower back stretch into your stretch routine is an effective way to promote flexibility.

Here are the steps:

- Lay on your back on your Yoga Strong Mat and hug your right knee into your chest. Let your right arm extend wide out to the right.
- Using your left hand, guide the right leg up and across the body to the left.
- Now, let your knee fall to the left. Bottom leg stays straight. Release your left arm to the left and if comfortable, bring your gaze to the right.
- Hold for 5 breaths. Repeat on the other side

Elbow-out Shoulder and Rotator Stretch: Stand with your hand in the back of the middle of your returned and your elbow mentioning. Reach over together with your different hand and gently pull your elbow ahead.

Rotating Forearm and Wrist Stretch: Place One arm directly out in the front and parallel to the floor. Rotate your wrist down and outwards and then use your different hand to further rotate your hand upwards.

5.3.6 Performance -

Performance sports aim at high sports performances and for that the physical and psychic capacities of sportsmen are developed to extreme limits. This normally does not happen in other areas of human activity. As a result, performance sports yield valuable knowledge about the limits to which human performance and various performance factors can be developed. It also leads to discovery of means and methods for improving various physical and psychic capacities (performance factors) to exceptionally high levels. This knowledge can be fruitfully applied to other areas of sports and human activity.

Variations in overall performance may also occur because of physiological, physical, tactical or psychological reasons (Marshall and Elliott, 1999). For instance, physiological factors together with fatigue, power and strength can impact on the way in which players perform physical skills, as can psychological factors which include stress and self-confidence. Poor anticipation and decision-making capabilities can equally affect players to make the incorrect tactical decision or to pick out an inappropriate method for the situation offered. Finally, maturational factors may also significantly affect the way in which technical talents are performed.

5.3.7 Sports Performance -

Sports performance is indeed an aspect of complex human performance which has several aspects or dimension several disciplines of sports science are required to work in a coordinated manner to explore the nature of sports performance and the process of improving sports performance. In the last few decades several disciplines of sports science have been established e.g., sports medicine, sports physiology, sports training, sports bio-mechanics, sports psychology, sports pedagogy, sports nutrition and so on.

Sports performance is the unity of execution and result of a sports action or a complex sequence of actions measured or evaluated according to socially determined and agreed norms. Sports performances are achieved in training or in competition.

5.3.8 Softball Performance -

Our softball performance training is delivered in a small group setting with a maximum of 8 athletes per session. Depending on the time of year, our sessions range from 45-75 minutes in length.

At the beginning of each session, we provide detailed demonstrations of each exercise to prepare our athletes to perform every exercise with the proper technique and know where they should feel each movement.

5.3.9 Importance of Softball -

Softball is a game wherein hundreds of thousands of members are engaged for fun. The two video games, fast pitch and gradual pitch, are difficult and exciting to play and thrilling to watch. Basically, they are very lots alike, the essential variations, as the name mean are located in the pace and trajectory because the pitch, the ball is constantly pitched underneath passed, but it is able to be lobbed throughout the of past due in a excessive arch are go speedy the better. The interesting action appeals to the Utah of the world, the competencies worried in the game are easy herbal and are incredibly stimulating and satisfying to anyone who participators in the game.

5.4 Purpose of the Study -

The research purpose is a statement of comparative analysis of effectiveness of stretching on performance of the softball players, the study is being conducted, or the goal of the study. The goal of a study might be to identify or describe a concept comparative analysis of effectiveness of stretching on performance of the softball players predict a situation or solution to a situation that indicates the type of study to be conducted performance of the softball players.

The game of softball is no exception. Softball is a very versatile game that makes enormous physical fitness, psychological, technical and tactical demands, but it is the physical fitness component that will directly determine the level of demand that can be placed on the technical, tactical, physiological and skills of a player.

5.5 Statement of the Problem –

A Comparative Analysis of Effectiveness of Stretching on Performance of the Softball Players.

5.6 Definition of Key Terms –

1. Stretching -

Stretching is a physical exercise that requires putting a body part in a certain position that'll serve in the lengthening and elongation of the muscle or muscle group and thus enhance its flexibility and elasticity.

2. Performance -

Performance of an individual is the individual efficiency in executing the skills and techniques in a fruitful manner during game situation which will be advantageous both for him-self and to the team for which he is playing. Performance will be assessed on 10 point rating scale; the rating will be based on subjective evaluation to predict the performance ability of softball players.

3. Softball -

Softball is a variant of baseball played with a larger ball on a smaller field. The game of softball originated as a kind of baseball for youngsters. Softball is a variant of baseball played with a larger ball on a smaller field. It was invented in 1887 in Chicago, Illinois, United States as an indoor game. It was at various times called indoor baseball, mush ball, playground, softball, kitten ball, and because it was also played by women, ladies' baseball. The name softball was given to the game in 1926, because the ball used to be soft.

5.7 Objectives -

The objectives formulated in the present study were as follows-

1. To find out the significant difference of effectiveness of stretching on performance of the softball players.
2. To find out the significant difference of effectiveness of stretching on performance of softball players of rural area.
3. To find out the significant difference of effectiveness of stretching on performance of softball players of urban area.

5.8 Hypotheses -

The hypotheses formulated in the present study were as follows-

1. There is no significant difference of effectiveness of stretching on performance of the softball players.
2. There is no significant difference of effectiveness of stretching on performance of softball players of rural area.
3. There is no significant difference of effectiveness of stretching on performance of softball players of urban area.
4. There is no significant difference of effectiveness of stretching on performance of boys' softball players.
5. There is no significant difference of effectiveness of stretching on performance of girls' softball players.

5.9 Delimitations -

The study was delimited in the following aspects-

1. Subjects of the present study were delimited to 170 softball players (Girls & Boys) at college level at Bikaner Region's rural and urban area.
2. Various stretching exercises are used for players.
3. The player's age ranged from 20 to 30 years.
4. The present study will be delimited to Softball players who had participated in Inter Collegiate and Zone tournaments.

5.10 Limitations -

The following factors are considered as limitations in the present study-

1. No special technique was used to motivate the subjects during the administration of the tests.
2. The basic of knowledge of the subjects in exercise science and their previous experience of physical activities were not taken into consideration.
3. Since the subjects were motivated verbally during testing and training periods no attempt was made to differentiate their level of motivation.
4. The heredity factors of the subjects and their influence on the selected variables were not taken into consideration.
5. Subject answers to statements in the questionnaire were influenced by a variety of variables, including the statements' level of knowledge, the seriousness and sincerity of the subjects, and so on.

5.11 Method -

Researcher has used **survey method** to get the information of the present circumstances.

5.12 Sample-

For the study researcher has selected one hundred seventy (N=170) male & female collegiate Softball players who were participated in the Inter Collegiate and inter-zone softball championship who represent respective colleges affiliated to various Universities in Rajasthan state and subject aged between 20 to 30 were selected as subjects for the study.

The full table of sample is shown given below-

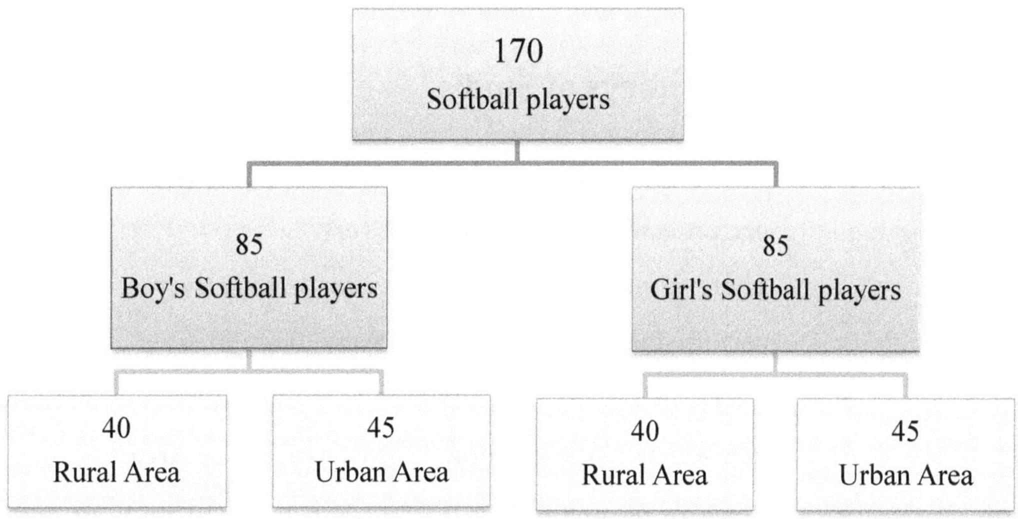

5.13 Tool used -

In this research the researcher used self-made questionnaire **'Effectiveness of Stretching on Performance of Softball Players'**. Here the researchers have used self-made questionnaire, this is-

- **Effectiveness of Stretching on Performance of Softball Players.**

5.14 Collection of data -

The data will be collected with self-made stretching effectiveness questionnaire. The data will be collected from Softball players who were participated in the Inter Collegiate and inter-zone softball championship of Rajasthan state. Researcher collects data from various colleges for raw score.

5.15 Statistical Techniques to be used -

Considering the aim of study and nature of data, following statistical techniques will be used for analysis-

1. Mean
2. Standard Deviation
3. T-test.

5.16 Major Findings of Study -

1. There is significant difference of effectiveness of stretching on performance of the Softball Players.
2. There is no significant difference of effectiveness of stretching on performance of the Softball Players of Rural Area.
3. There is significant difference of effectiveness of stretching on performance of the Softball Players of Urban Area.
4. There is no significant difference of effectiveness of stretching on performance of boys' Softball Players.
5. There is no significant difference of effectiveness of stretching on performance of girls' Softball Players.

5.17 Conclusion -

The current research work was undertaken in the effectiveness of stretching on performance of softball players who were participated in the Inter Collegiate and inter-zone softball championship of Rajasthan state.

The result showed that find significant difference between Boys players & Girls players. The performance Score of Girls are high than Boys. The result showed that the urban Girls players' stretching effectiveness high than urban Boys players.

5.18 Suggestions for Further Research -

After conducting this study, the researcher found many other facts about the related study. The researcher wishes to recommend this study for further investigation on the basis of following aspects. Based on the problem of the study, the following suggestions are made.

- Its benefit, if we give more focus on psychological variables of softball players.
- In this study only inter collegiate players were selected; in future we can select school board level players.
- Similar study can be taken up on different gender, age group, level of participation.
- Further research may be conducted to assess only performance and related variables of 400 or more than players.
- This study was done only one State (Rajasthan), in future we can do this research two or more state's player.
- Also doing this research with same variables as a comparison study of two or more state's players.
- Also doing this research with different variables as a comparison study of two or more state's players of different games.
- Further research may be conducted same variable on the other games' players.
- After 5-6 year later we can be conducted same research on present situations.

REFERENCES

1. **Abd Rahim Bin Mohd Sharif et al (2014),** The Impact of Hockey Coaches and Team Cohesion on the Performance of Players, Volume No.12,Issue No.4,P No.1-13, ISSN 1911-2025.
2. **Abraham, G. (2010),** Analysis of anthropometry, body composition and performance variables of young Indian athletes in southern region. Indian Journal of Science and Technology, 3(12):1210-1213.
3. **Adam, J.J. and Wilberg (1992),** "Individual differences in Visual informative processing rate and the prediction of performance differences in team sports – A Preliminary Investigation", Journal of Sports Science, 28: 261-273.
4. **Aggarwal. J.C (1975),** Educational Research, New Delhi, Aryan Book Depot, 1975, P. 73.
5. **Alderman, R.B. (1974),** "Psychological Behaviour in Sports" London: W. B Saunders Company.
6. **Akbar, S. K., & Khan, D. J. A. (2019),** A factor structure study on anthropometric and physical fitness variables of softball players. International Journal of Yogic, Human Movement and Sports Sciences, 4(1), 794–800.
7. **American Alliance for Health, Physical Education and Recreation (1991),** Softball skills test manual for boys and girls. Roberta Rikli (Ed.), AAHPER, Reston, VA, USA.
8. **Amiri-Khorasani M, Kellis E (2013),** Static vs. Dynamic acute stretching effect on quadriceps muscle activity during soccer instep kicking. J Hum Kinet. 39:37–47.
9. **Amy Marquardt (2018),** Effects of Starting Stance on Base Running Sprint Speed in Softball Players, Volume No.1, Issue No.8, PP No.1-8.
10. **Aswathy, V. (2018),** The Relation between Hand Grip Strength with Hand-Anthropometric Variable in Inter-University Level Softball, Cricket and Hockey Players. International Journal of Recent Scientific Research, 9(4(H)).
11. **Arzola, G. (2006),** Throwing a softball fast. Coach and Athletic Director, 75, 46-8.

12. **Bale P. (1986),** A review of the physique and performance qualities characteristics of games players in specific positions on the field of play, The Journal of sports medicine and physical fitness (quarterly review) Vol. 26, No.2, P 109 – 122.
13. **Behm DG, Chaouachi A. (2011),** A review of the acute effects of static and dynamic stretching on performance. Eur J Appl Physiol.;111(11):2633–2651.
14. **Behm DG, Plewe S, Grage P, et al. (2011),** Relative static stretch-induced impairments and dynamic stretch-induced enhancements are similar in young and middle-aged men. Appl Physiol Nutr Metab. ; 36(6):790–797.
15. **Bhagat, U., Singh, A. and Deol, N.S. (2015),** Comparative study of selected anthropometric, physical fitness and psychological variables between softball and cricket state level boys players. Indian Journal of Applied Research, 5(6):557-560.
16. **Bishop, D.J. and Girard, O. (2013),** Determinants of team-sport performance: implications for altitude training by team-sport athletes. British Journal of Sports Medicine, 47(Suppl 1):i17-21.
17. **Boden, L. M., Rosengren, K. J., Martin, D. F., & Boden, S. D. (2009),** A comparison of static near stereo acuity in youth baseball / softball players and non–ball players. Optometry - Journal of the American Optometric Association, 80(3), 121–125.
18. **Bracko MR. (2002),** Can stretching prior to exercise and sports improve performance and prevent injury? ACSM's Health & Fitness Journal; 6(5).
19. **Brule R. Etnyre and Eva J. Lec. (1988),** "Chronic and Arm Flexibility of men & women using three different stretching techniques. Research Quarterly for Exercise & sports, Vol. 59, No.3, P222.
20. **Brylinskyl, J. and Melanie, M.J.C.F. (1992),** The effect of using a weighted softball on pitching velocity, wrist strength and handgrip. Journal of Strength and Conditioning Research, 6(3):170-173.
21. **Burris, K., Vittetoe, K., Ramger, B., Suresh, S., Tokdar, S. T., Reiter, J. P., & Appelbaum, L. G. (2018),** Sensorimotor abilities predict on-field performance in professional baseball. Scientific Reports, 8(1), 116.
22. **Carter, J.E.L. (1985),** Morphological factors limiting human performance, in limits of human performance (eds D. Clarke and H. Eckert). American

academy of physical education papers no. 18, Human Kinetics, Champaign, IL, pp. 106-107.

23. **Chang, Y. K., Ho, L. A., Lu, F. J. H., Ou, C. C., Song, T. F., & Gill, D. L. (2014),** Self-talk and softball performance: The role of self-talk nature, motor task characteristics, and self-efficacy in novice softball players. Psychology of Sport and Exercise, 15(1), 139–145.

24. **Chaouachi, Anis & Castagna, Carlo & Chtara, Moktar & Brughelli, Matt & Turki, Olfa & Galy, Olivier & Chamari, Karim & Behm, David. (2009),** Effect of Warm-Ups Involving Static or Dynamic Stretching on Agility, Sprinting, and Jumping Performance in Trained Individuals. Journal of strength and conditioning research / National Strength & Conditioning Association. 24. 2001-11. 10.1519/JSC.0b013e3181aeb181.

25. **Chatzopoulos D, Galazoulas C, Patikas D, et al.(2014),** Acute effects of static and dynamic stretching on balance, agility, reaction time and movement time. J Sports Sci Med. 2014; 13(2):403–409.

26. **Chtourou H, Aloui A, Hammouda O, et al.(2013),** Effect of static and dynamic stretching on the diurnal variations of jump performance in soccer players. PLOS ONE. 2013; 8(8):e70534.

27. **Craig, S.B., Craig, S. and Johnson. K. (1985),** The softball handbook. Leisure press, p. 16-22.

28. **Daksh Sharma (2017),** A comparative study of anthropometric variables of female softball players with non-softball players, Volume No.4, Issue No.5, PP No. 1-2.

29. **Das, S.S., (1995),** Effect of two specific training programmes on the performance of young Soccer player, unpublished Doctoral thesis, University of Kalyani.

30. **Dass, S.S. and Banarjee, A.K (1992),** Variation in duration of training period on the performance variables of young soccer players. NIS Scientific Journal 15:3, P116 – 21.

31. **Dickson, N.L., Ruot, C., Madeson, M. and Lindsay Edwards, L. (2009),** A descriptive and comparative study of physical and performance characteristics of NCAA Division II and Division III softball players. International Journal of Exercise Science, 2(1):S30-31.

32. **Douda, H.T., Toubekis, A.G., Avloniti, A.A. and Tokmakidis, S.P. (2008),** Physiological and anthropometric determinants of rhythmic gymnastics performance. International Journal of Sports Physiology and Performance, 3:41-54.

33. **Dr. Nasreen Ashra et al (2020),** The Immediate Effect of Sleeper Stretch and Proprioceptive Neuromuscular Facilitation Stretch on Shoulder Range of Motion in Badminton Player, Volume No.19,PIsssue No.10,PP No.1-6, ISSN: 2279-0861.

34. **Elferink-Gemser, M.T., Visscher, Ch., Lemmink, K.A. and Mulder, T.W. (2004),** Relation between multidimensional performance characteristics and level of performance in talented youth field hockey players. Journal of Sports Sciences, 22:1053-1063.

35. **Ellis, L. (2000),** Protocols for the physiological assessment of softball players. Physiological Tests for Elite Athletes, Ch.26, 363-71.

36. **F. Spaniol, A Quinonez, S Cochran, B Hicks, M Alves, & B Warren. (2014),** The Relationship Between Visual Skills And Batting Performance of Professional Baseball Players.

37. **Faigenbaum, A. D., Bellucci, M., Bernieri, A., Bakker, B., & Hoorens, K. (2005),** Acute effects of different warm-up protocols on fitness performance in children. The Journal of Strength & Conditioning Research, 19(2), 376-381.

38. **Fajar Awang Irawan (2015),** The Effectiveness of Sidearm Throw in Softball Players,Volume No.5,Issue No.2,PP No.1-7, ISSN: 2088-6802.

39. **Feliu Rovira, A. (1991),** "Prediction of physical endurance in athletes during puberty: analysis of high performance soccer players", An Esp. Pediatr. 35(5): 323-6.

40. **Feldhacker, D. R., Molitor, W. L., Athmann, A., Boell, M., Kaiser, A., Musch, A., &Willhite, L. (2019),** Efficacy of High-performance Vision Training on Improving the Reaction Time of Collegiate Softball Athletes: A Randomized Trial. Journal of Sports Medicine and Allied Health Sciences: Official Journal of the Ohio Athletic Trainers' Association, 4(3), 1–12.

41. **Fletcher and Jones (2004),** The effect of different warm-up stretch protocols on 20 meter sprint performance in trained rugby union players, PMID:15574098, Nov;18(4):885-8.

42. **Fletcher IM, Monte-Colombo MM (2010),** An investigation into the possible physiological mechanisms associated with changes in performance related to acute responses to different pre activity stretch modalities. The Journal of Applied Physiology. NRC Research Press (Canada) Feb; 35(1):P27-34.
43. **Flyger, N., Button, C. and Rishiraj, N. (2006),** The science of softball: implications for performance and injury prevention. Sports Medicine, 36(9):797-816.
44. **Gabbett, T.J. and Seibold, A.J. (2013),** Relationship between tests of physical qualities, team selection, and physical match performance in semiprofessional rugby league players. Journal of Strength and Conditioning Research, 27(12):3259-3265.
45. **Gabbett, T.J., Jenkins, D.G. and Abernethy, B. (2011),** Relationships between physiological, anthropometric, and skill qualities and playing performance in professional rugby league players. Journal of Sports Sciences, 29(15):1655-1664.
46. **Gabbett, T., Rubinoff, M., Thorburn, L., & Farrow, D. (2007),** Testing and Training Anticipation Skills in Softball Fielders. International Journal of Sports Science & Coaching, 2(1), 15–24.
47. **Goswami, N. (2014),** Prediction of performance ability of sprinters, jumpers and throwers in relation to selected motor fitness components and physiological variables. International Journal of Sports Science, Fitness and Leisure Industry, 1(1):86-98.
48. **Hallvard Nygaard Falch et al (2019),** Effect of Different Physical Training Forms on Change of Direction Ability: a Systematic Review and Meta-analysis, Volume No.1, issue No.9, PP No.1-37.
49. **Herda, T. J., Cramer, J. T., Ryan, E. D., McHugh, M. P., & Stout, J. R. (2008),** Acute effects of static versus dynamic stretching on isometric peak torque, electromyography, and mechanomyography of the biceps femoris muscle. The Journal of Strength & Conditioning Research, 22(3), 809-817.
50. **Hitoshi Shitara et al (2016),** Shoulder Stretching Intervention Reduces the Incidence of Shoulder and Elbow Injuries in High School Baseball Players: a Time-to-Event Analysis, Volume No.2, Issue No.2, PP No.1-6.

51. **Hoffman JR, Vazquez J, Pichardo N, Tenenbaum G, (2009),** Anthropometric and performance comparisons in professional baseball players. Journal of Strength and Conditioning Research; 23(8):2173-2178.

52. **Houtkooper, L.B. and Going, S.B. (1994),** Body composition: How should it be measured? Does it affect sport performance? Sports Science Exchange 7: supplement 52.

53. **Ibrahim Hamed Ibrahim Hassan (2018),** Relationship between strength, speed and change direction performance in field hockey players, Volume nO.2,Issue No.1,PP No.1-5.

54. **Ikeda, Mariko (2004),** "Physique and Motor Performance of College Women Handball Players", https://qir.kyushu-u.ac.jp/dspace/bitstream/2324/10775/1/p033.pdf.

55. **Israel, R. G., & Brown, R. L. (1981),** Response Time Comparisons Among Four Base Running Starting Techniques in Slow Pitch Softball. Research Quarterly for Exercise and Sport, 52(3), 324–329.

56. **J. Matt Mccrary et al (2015),** A systematic review of the effects of upper body warm-up on performance and injury, Volume No.1, Issue No.5, PP No.1-9.

57. **Jodie Darling, Mike Riggenbach, Sean P. Flanagan and Lloyd L and Laubach. (2001),** The University of Dayton. "The Effects of different strength training modes on performance in selected motor activities for prepubescent. Research Quarterly, Vol.72 No – 1, A – 20.

58. **Jones B. J., Murray M. J. (1978),** Softball concepts for coaches and teachers, WMC Brown. p- 2&7.

59. **Jung-Hyun Choi (2019),** Effects of Elastic Band Exercise Using Proprioceptive Neuromuscular Facilitation on Strength and Dynamic Balance of Adults with Ankle Instability, Volume No.1,Issue No.8,PP No.1-7.

60. **Kakran, S. S. (2016),** A comparative study on coordinative abilities among male softball players and cricketers. International Journal of Physical Education, Sports and Health, 3(1), 34–36.

61. **Kakran, S. S., & Gurjar, K. S. (2016),** A comparative analysis on motor fitness between university level cricket and softball players. International Journal of Physical Education, Sports and Health, 3(1), 153–154.

62. **Kanwar Mandeep Singh et al (2017),** relationship between the anthropometric variables and throwing skill in male softball players, Volume No.3,Isue No.10,PP NO.1-11, ISSN: 2501 – 1235.

63. **Kanwar Mandeep Singh (2017),** Relationship of physiological parameters with performance among softball players, Volume No.2,issue NO.5,PP No.1-4, ISSN: 2455-4197.

64. **Kanwar Mandeep Singh (2017),** Relationship of physiological parameters with the fielding skill among the male softball players, Volume No.2,Issue No.5,PP NO.1-4, ISSN: 2456-5067.

65. **Kanwar Mandeep Singh (2017),** Association of physiological parameters with the throwing performance among the male softball players, Volume No.2,Issue No.2,PP NO.1-4., ISSN: 2456-4419.

66. **Kaur, S., & Bhagat, U. (2016),** Assessment of reaction time and steadiness between Inter-College and Inter-University level female softball Players. International Journal of Physical Education, Sports and Health, 3(5), 92–94.

67. **Kelly P. Massey (2018),** A comparison of base running start techniques in collegiate fast pitch softball athletes, Volume No.1, Issue No.3, PP No.1-7.

68. **Knechtle, B., Duff, B., Amtmann, G. and Kohler, G. (2007),** Cycling and running performance, not anthropometric factors, are associated with race performance in a triple iron triathlon. Research in Sports Medicine, 15:257-269.

69. **Kobak, M. S., Rebold, M. J., Buser, S. L., Kappler, R. M., & Otterstetter, R. (2018).** The Effects of Medicine Ball Training on Bat Swing Velocity in Prepubescent Softball Players. International Journal of Exercise Science, 11(4), 75–83.

70. **Koley, S. and Kumaar, S.B. (2012),** The relation between handgrip strength and selected hand-anthropometric variables in Indian inter-university softball players. Facta Universitatis series: Physical Education and Sport, 10(1):13-21.

71. **Korhonen, M.T. (2003),** "Age-related differences in 100-m sprint performance in male and female master runners", Med. Sci. Sports Exerc. 35(8): 1419-28.

72. **Kovacs, M. (2009),** Dynamic stretching: The revolutionary new warm-up method to improve power, performance and range of motion. Ulysses Press.

73. **Koya Mine et al (2017),** Effectiveness of Stretching on Posterior Shoulder Tightness and Glenohumeral Internal-Rotation Deficit: A Systematic Revie of Randomized Controlled Trials, Volume No.1, Issue No.7.PP No.1-13.

74. **Krane, V., Joyce, D., & Rafeld, J. (2016),** Competitive Anxiety, Situation Criticality and Softball Performance. The Sport Psychologist, 8(1), 58–72.

75. **Kumar, D. and Singh, B. (2014),** Comparative study of flexibility and leg strength between korfball and softball women players of Delhi. International Journal of Multidisciplinary Research and Development, 1(6):64-66.

76. **Kumaresan P (2003),** Influence of plyometric Training and Aerobic Training and Anaerobic Power, Aerobic Power and skill performance among soccer players. ASIA SPECIFIC, international conference for sports administers souvenir.

77. **Lane B. Bailey et al (2017),** Effectiveness of Manual Therapy and Stretching for Baseball Players With Shoulder Range of Motion Deficits, Volume No.1,Issue No.3,PP No.1-8.

78. **Larry W. Titlow, Billy Mark Young, and Kevin H. Kendrick (2006),** University of Central Arkansas, acute effects of stretching on knee flexion. Research Quarterly, March supplements Vol. 77 No.1, P A–21–22.

79. **Little T, Williams AG. (2006),** Effects of differential stretching protocols during warm-ups on high-speed motor capacities in professional soccer players. J Strength Cond Res.; 20(1):203-307.

80. **Lloyd Smith (2008),** Progress in measuring the performance of baseball and softball bats, Volume No.1, Issue No.6, PP No.1-9.

81. **Lowe, H.E., Szymanski, D.J., Bankston, B.L., Braswell, M.T., Britt, A.T., Gilliam, S.T., Herring, A.L., Holloway, B.T., Lowe, D.W., Potts, J.D., Szymanski, J.M. and Till, M.E. (2010),** Relationship between body composition and bat swing velocity of college softball players. Journal of Strength and Conditioning Research, 24(P1).

82. **Mandeep Singh et al (2019),** Study of body composition among male and female softball players, Volume No.5,Issue No.10.PP NO.109, ISSN: 2501 – 1235.

83. **Mangaonkar Anagha, Putambekar Anjali (Decembert 2018),** Effect of suryanamaskar vs. dynamic stretching on hamstring flexibility among

physiotherapy students. Int J Physiother. Vol 5 (6), 202-206, ISSN: 2348-8336.

84. **Manoel ME, Harris-Love MO, Danoff JV, Miller TA. (2008),** Acute effects of static, dynamic, and proprioceptive neuromuscular facilitation stretching on muscle power in women. J Strength Cond Res.; 22(5):1528-1534.

85. **Magrini, M., Dawes, J. J., Spaniol, F. J., & Roberts, A. (2018),** Speed and Agility Training for Baseball/Softball: Strength and Conditioning Journal, 40(1), 68–74.

86. **Marquardt, A., Wong, M. A., Watkins, C. M., Barillas, S. R., Galpin, A. J., Coburn, J. W., & Brown, L. E. (2018),** Effects of Starting Stance on Base Running Sprint Speed in Softball Players. International Journal of Exercise Science, 11(6), 179–186.

87. **McMillian DJ, Moore JH, Hatler BS, et al. (2006),** Dynamic vs. static-stretching warm up: the effect on power and agility performance. J Strength Cond Res.; 20(3):492–499.

88. **McMillian DJ1, Moore JH, Hatler BS, Taylor DC. (2006),** Dynamic vs. static-stretching warm up: the effect on power and agility performance. J Strength Cond Res. Aug; 20(3):492-9.

89. **Md Rahim, N. I. Bin, & Krasilshchikov, O. (2015),** The Effects of Pre-Season Resistance Training Program on Fitness Profile of Female University Softball Players. International Journal of Physical Education, Fitness and Sports, 4(4), 53–58.

90. **Meswaniya, N.K. (2012),** Comparison of selected physical fitness variables of school level softball and cricket players. Indian Journal of Research, 1(9):168-169.

91. **Miller, R. G., & Shay, C. T. (1964),** Relationship of Reaction Time to the Speed of a Softball. Research Quarterly. American Association for Health, Physical Education and Recreation, 35(3), 433–437.

92. **Moller MH, Oberg BE and Gillquist J (1985),** Stretching exercise and soccer: effect of stretching on range of motion in the lower extremity in connection with soccer training. International Journal of Sports Medicine 1985 Feb;6(1):50-2. Georg Thieme Verlag Pub. (US)

93. **Nathan J. Hoffmeier (2014),** Effects of a single-session graston technique on hamstring flexibility and muscle stiffness of recreationally active individuals, Volume No.1, Issue No.4, PP No.1-14.

94. **Nicholas J. Kelling (2018),** The effect of spatial working memory capacity on ball flight perception, Volume No.1, Issue No.4, PP No.1-14.

95. **Nimphius, S. (2010),** Lag time: the effect of a two week cessation from resistance training on force, velocity and power in elite softball players. Journal of Strength and Conditioning Research, 24(P1).

96. **Nimphius, S., McGuigan, M.R. and Newton, R.U. (2012),** Changes in muscle architecture and performance during a competitive season in female softball players. Journal of Strength and Conditioning Research, 26(10):2655-2666.

97. **Nimphius, S., McGuigan, M.R., and Newton, R.U. (2010),** Relationship between strength, power, speed and change of direction performance of female softball players. Journal of Strength and Conditioning Research, 24(4):885-895.

98. **Novotna, S. (2010),** Fitness development model of female softball players. Joint International IGIP-SEFI Annual Conference, 19th-22nd September 2010, Trnava, Slovakia.

99. **Nuri, L., Shadmehr, A., Ghotbi, N., & Attarbashi Moghadam, B. (2013),** Reaction time and anticipatory skill of athletes in open and closed skill dominated sport. European Journal of Sport Science, 13(5), 431–436.

100. **Olfa Turki, Wissem Dhahbi, Johnny Padulo, Riadh Khalifa, Sana Ridène, Khaled Alamri, Sabri Gueid, Karim Chamari (2019),** Warm-Up With Dynamic Stretching: Positive Effects on Match-Measured Change of Direction Performance in Young Elite Volleyball Players.

101. **Özdemir, R.A.., Kirazcı S., Ugras, A. (2010),** Simple reaction time and decision making performance after different physical workloads: an examination with elite athletes. International Journal of Human Sciences Volume 7, Issue 2.

102. **Pardeshi & Kadam. (2016),** A comparative study of physical fitness components between Baseball and Softball players of Pune City. International Journal of Physiology, Nutrition and Physical Education 2016; 1(2): 218-219.

103. **Paul Colosky, David Newton, and Vinson H. Sutlive, (2001),** Research Quarterly, March, Vol.72 No2, Review of Literature on Stretching and Strength Performance. Colorado State University.

104. **Peart, A., Wadsworth, D., Washington, J., & Oliver, G. (2018),** Body Composition Assessment in Female National Collegiate Athletic Association Division I Softball Athletes as a Function of Playing Position Across a Multiyear Time Frame. Journal of Strength and Conditioning Research, 33(11), 1–7.

105. **Peck, Evan MD; Chomko, Greg DPT; Gaz, Dan V. MS; Farrell, Ann M. MLS (2014),** The Effects of Stretching on Performance Current Sports Medicine Reports: Volume 13 – Issue 3 - p 179-185.

106. **Poltavski, D., & Biberdorf, D. (2015),** The role of visual perception measures used in sports vision programmes in predicting actual game performance in Division I collegiate hockey players. Journal of Sports Sciences, 33(6), 597–608.

107. **Pushkar Yadav (2014),** A study of selected biomechanical variables as a factor of hitting performance among softball players, Volume No.1, Issue No.4,PP No.1-3, ISSN: 2347-6745.

108. **Rafael E. Reigal (2019),** Psychological Profile, Competitive Anxiety, Moods and Self-Efficacy in Beach Handball Players, Volume No.1, Issue No.3,PP No.1-13.

109. **Rao, S.K. and Kumar, P.P.S. (2015),** The relationship between selected physiological parameters variables with playing ability of softball players. International Journal of Engineering Research and Sports Science, 2(4):1-4.

110. **Rodney J. Negrete et al (2012),** can upper extremity functional tests predict the softball throw for distance: a predictive validity investigation, Volume No.6, Issue No.2, PP No.1-8.

111. **Rokhayah, F., Kristiyanto, A., & Sugiyanto. (2017),** Influence of the Difference of Perception and Kinesthetic Exercise Methods against Precision Hit the Ball Softball. Journal of Physical Education Health and Sport, 4(1), 23–28.

112. **Rozella Ab. Razak et al (2019),** Combined Strength Training on Throwing Ball Velocity in Softball, Volume No.1,Issue NO.2,PP No.1-12.

113. **Rubini EC, Costa AL, Gomes PS (2007),** The effects of stretching on strength performance. Sports Med; 37(3):213–224.

114. **Ryan r. Fairall et al (2017),** Acute effects of self-myofascial release and stretching in overhead athletes with GIRD, Volume No.1,Issue No.7,PP No.1-6.

115. **S. Sarika Waghmare Sumedhkumar Balajirao, Shweta Shenoy (2019),** Evaluation Of Acute Effects Of Combined Stretching Methods On Flexibility, Agility And Speed Among Cricket Players, European Journal of Physical Education and Sport Science, ISSN: 2501 - 1235 ISSN-L: 2501 – 1235. Open Access Publishing Group 29 doi: 10.5281/zenodo.3247882 Volume 5 | Issue 10.

116. **Sakurai, S. and Ohtsuki, T. (2000),** "Muscle activity and accuracy of performance of the smash stroke in badminton with reference to skill and practice." J. Sports Sci. 18(11), pp.901-14.

117. **Sally J. Ford (2009),** Implementing a Breathing Technique to Manage Performance Anxiety in Softball, Volume No.2, Issue No.3, PP No.1-9.

118. **Sang-Won Lee et al (2016),** The Effect of PNF Stretching with Elastic Band on Ball Speed of High School Baseball Players, Volume 14, No 12, PP 525-535, ISSN: 1738-1916.

119. **Saraya, A. E., Sugiyanto, & Doewes, M. (2018),** Anthropometric Factors and Physical Condition Dominant Determinants Batting Skills In Softball. International Journal of Multicultural and Multi religious Understanding, 5(4), 213–222.

120. **Sathuluri, R., Rao, K. S., & Kumar, P. P. S. P. (2016),** The Relationship between Selected Motor Fitness Anthropometric Parameters with Playing Ability of Softball Players: International Journal of Engineering Research and Sports Science, 3(5), 83–87.

121. **Sayers AL, Farley RS, Fuller DK, Jubenville CB and Caputo JL (2008),** The effect of static stretching on phases of sprint performance in elite soccer players. The Journal of Strength and Conditioning Research (JSCR).2008 Sep;22(5):1416-21 .

122. **Sharma, V., Kumar, S., Singh, G., Kumar, P. and Chandel, R. (2014),** A comparative study of selected physiological and physical parameters between inter-university and inter-college male softball players of Punjab. Proceedings

of Global Scientific Conference on Physical Education, Health and Sports Sciences (GLOPEHSS), Department of Physical Education (Teaching), Guru Nanak Dev University, Amritsar, 2:51-54.

123. **Shashikant Pardeshi and Dr. Govind Kadam (2016),** A comparative study of physical fitness components between Baseball and Softball players of Pune City, Volume No.1,Issue No.2, PP No.1-2, ISSN: 2456-0057.

124. **ShibNath JHA (1984),** Effect of metabolic warm-up specific and general on soccer skill performance, M.P.E. 1979. (P.J. Sebestian). Thesis Abstracts, P. 105, (1984). L.N.C.P.E. Gwalior.

125. **Shrier I. (2004),** Does stretching improve performance? A systematic and critical review of the literature. Clin J Sport Med.; 14(5):267–273.

126. **Shrikant S Mankar (2020),** The role of yogic exercises on softball players, Volume No.5,Issue No.2,PP No.-1-2.

127. **Shyamal Koley (2011),** Correlations of handgrip strength with selected hand-anthropometric variables in university softball players, Volume No.3, Issue No.1, PP No.1-5.

128. **Singh, A. and Gaurav, V. (2014),** Comparative study of hand grip and shoulder girdle strength among inter college level cricket, baseball and softball players. International Multidisciplinary Research Journal, 2(3):1-5.

129. **Singh, K. M. (2017),** Relationship of physiological parameters with performance among softball players. International Journal of Academic Research and Development, 2(5), 602-605.

130. **Singh, K. M., Singh, M., Singh, P., & Choundhary, A. (2017),** Relationship between the Anthropometric Variables and Throwing Skill in Male Softball Players. European Journal of Physical Education and Sport Science, 3(10), 198-208.

131. **Singh, M. (2018),** Association of Anthropometric Characteristics and Body Composition with the Performance among Male Softball Players. European Journal of Physical Education and Sport Science. 4(3), 13–31.

132. **Singh, P. (2014),** Comparative study of physical fitness and psychological variables of softball and cricket players. Online International Interdisciplinary Research Journal, 4(2):246-251.

133. **Singh, S., & Singh, M. (2015),** Relationship of Anthropometric Characteristics with Batting Skill in Softball among Male Softball Players.

International Journal of Sports Science, Fitness and Lesuire Industry, 1(4), 38–52.

134. **Singh, S., Singh, M. and Rathi, B. (2013),** Kin anthropometric and performance characteristics of elite and non-elite female softball players. Journal of Sports Medicine and Physical Fitness, 53(6):628-634.

135. **Sintara, K. and Sonchan, N. (2015),** Physical fitness of collegiate softball players, Burapha University, Thailand. Proceedings of the Burapha University International Conference, 10-12 July 2015, Bangsaen, Chonburi, Thailand.

136. **Sobolewski EJ, Wagner DR, Bressel E. (2019),** Effect of static stretching and jogging on knee extension isokinetic peak torque. Isokinetics & Exercise Science; 19(3):157–162.

137. **Sophia Nimphius et al (2010)**, relationship between strength, power, speed, and change of direction performance of female softball players, Volume No.1,Issue No.9,PP No.1-11.

138. **Szymanski, D.J., Bassett, K.E., Beiser, E.J., Till, M.E., Medlin, G.L., Beam, J.R. and Derenne, C. (2012),** Effect of various warm-up devices on bat velocity of intercollegiate softball players. Journal of Strength and Conditioning Research, 26(1):199-205.

139. **Taleb-Beydokhti, I., & haghshenas, rouhollah. (2015),** Static versus dynamic stretching: Chronic and acute effects on Agility performance in male athletes. International Journal of Applied Exercise Physiology, 4(1), 1-8.

140. **Terbizan DJ, Waldera M, Seljevold P, Schweigert DJ, (1996),** Physiological characteristics of masters' women fastpitch softball players. Journal of Strength and Conditioning Research; 10:157-160.

141. **Timothy J. Suchomel et al (2013),** a comparison of baseball and softball players' bilateral strength asymmetry and its relationship with performance, Volume No.1, issue No.3, PP No.1-37.

142. **Thompson, K.J. and Barkley, S.A. (2013),** Evaluation of strength and conditioning changes during a Division II collegiate softball session. Journal of Fitness Research, 2:55-63.

143. **Till, M.E., Bassett, K.E., Beiser, E.J., Medlin, G.L., Szymanski, J.M., Brooks, K.A. and Szymanski, D.J. (2011),** Relationship between lower body power, body mass, and softball-specific skills. Journal of Strength and Conditioning Research, 25.

144. **Troumbley, Patrick (2010),** Static Versus Dynamic Stretching Effect on Agility Performance. All Graduate Theses and Dissertations.

145. **Turki, O., Chaouachi, A., Drinkwater, E. J., Chtara, M., Chamari, K., Amri, M., & Behm, D. G. (2011),** Ten minutes of dynamic stretching is sufficient to potentiate vertical jump performance characteristics. The Journal of Strength & Conditioning Research, 25(9), 2453-2463.

146. **Upma Bhagat et al (2015),** Comparative Study of Selected Anthropometric, Physical Fitness and Psychological Variables between Softball and Cricket State Level Boys Players, Volume No5, Issue No.6, PP No.1-4, ISSN - 2249-555X.

147. **Vinson H. Sutlive (2001),** Research Quarterly, March, Vol.72 No1, (A16) Review of Literature on Stretching and Strength Performance. Colorodo State University.

148. **Warren Young and Simon Elliott, (2001),** Acute effects of static stretching, proprioceptive neuromuscular facilitation stretching and maximum voluntary contractions on explosive force production and jumping performance, Research Quarterly for exercise and sport, Sep, Vol. 72, No.3, P 273 – 278.

149. **W.B.Young and D.G.Behm (2003),** Effects of running , static and stretching and practice jumps on explosive force production and jumping performance, The Journal of Sports Medicine and Physical Fitness 43(1):21-7.

150. **Weerapong P, Hume P, Kolt G (2004),** Stretching: mechanisms and benefits for sport performance and injury prevention. Physical Therapy Reviews.; 9(4):189.

151. **Wentzel, M. and Travill, A.L. (2015),** Relationship among fitness, morphological characteristics, skills and performance in men's fast-pitch softball. South African Journal for Research in Sports, Physical Education and Recreation, 37(2):175-186.

152. **Werner, S.L., Deryk, G.J., Guido, J.A. and Brunet, M.E. (2006),** Kinematics and kinetics of elite windmill softball pitching. American Journal of Sports Medicine, 34:597-603.

153. **Withers, R.T. and Robert, R.G.D. (1981),** Physiological profiles of representative women softball, hockey and netball players. Ergonomics, 24(8):583-591.

154. **Wong PL, Lau PW, Mao de W, Wu YY, Behm DG and Wisloff U (2011),** Three days of static stretching within a warm-up does not affect repeated-sprint ability in youth soccer players. The Journal of Strength and Conditioning Research (JSCR). Mar;25(3):838-45.

155. **Yamaguchi, T., & Ishii, K. (2005),** Effects of static stretching for 30 seconds and dynamic stretching on leg extension power. The Journal of Strength and Conditioning Research, 19(3),

156. **Yanagisawa, O., Maegawa, T. and Funato, K. (2009),** The anthropometric characteristics of Japanese baseball and female softball players. Gazzetta Medica Italiana Archivio per le Scienze Mediche,168(1):11-21.

157. **Young W, Clothier P, Otago L, Bruce L, Liddell D (2004),** Acute effects of static stretching on hip flexor and quadriceps flexibility, range of motion foot speed in kicking a football. The Journal of Strength and Conditioning Research (JSCR). 2004 Mar;7(1): P 23-31.

158. **Zakas A, Grammatikopoulou MG, Zakas N, Zahariadis P and Vamvakoudis E (2006),** The effect of active warm-up and stretching on the flexibility of adolescent soccer players. The Journal on Applied Physiology, Preventive Medicine, Sports Medicine and Traumatology, Sports Psychology. 2006 Mar;46(1):P 57-61. Edizioni Minerva Medica Pub, (Italy).

Website:

- kmy/upjss/june2011
- https://en.wikipedia.org/wiki/Warming_up
- http://www.rnarayanaswami.net/yogapdf
- gudetoptpractice.apta.org
- https://orsupply.com/catalog/documents/aboutPulseOximetry.pdf
- https://en.wikipedia.org/wiki/Agility
- https://en.wikipedia.org/wiki/Physical_strength
- https://en.wikipedia.org/wiki/Blood_pressure
- https://www.amperordirect.com/pc/help-pulse-oximeter/z-what-isoxygen- saturation.html
- www.softball.com/
- www.asasoftball.com

www.ingramcontent.com/pod-product-compliance
Lightning Source LLC
LaVergne TN
LVHW082337200926
848481LV00012B/3375
* 9 7 9 8 2 3 1 2 1 9 4 6 9 *